In Pursuit of Hope
A Guide for the Seeker

Dedicated

to youth everywhere
who will pass on to their children
the world they want to build

In Pursuit of Hope
A Guide for the Seeker

Arthur Lyon Dahl

GEORGE RONALD • OXFORD

George Ronald, Publisher
Oxford
www.grbooks.com

A catalogue record for this book is available from the British Library

ISBN 978-0-85398-620-1

Cover design: René Steiner, Steinergraphics.com

CONTENTS

Preface

Hope has become a rare commodity in today's world. Everything seems to be going wrong, and the forces of disintegration are accelerating. We may wonder what will come first: a financial collapse, a climate catastrophe, a global pandemic, a third world war, or some equally disastrous outcome for a material civilisation out of control? The youth of the world in particular face dismal prospects. Yet there are also forces of integration at work; people making a difference and living lives of meaning. If there were more of them, that might help us to avoid the worst and turn the corner towards a brighter tomorrow.

In my professional work as an environmental scientist, I have spent over half a century researching and spreading depressing news about the environmental crises on the horizon or already happening. So I asked myself what could I do to compensate and give some hope for the future? Why did I succeed in staying hopeful and motivated despite so much evidence to the contrary? Surprising as it may seem for a natural scientist, especially one specialising in complex ecosystems like coral reefs now threatened with disappearance, the answer for me was my perspective as a member of the Bahá'í Faith, founded in the 19th century by the Báb and Bahá'u'lláh, and anticipating the challenges we now face. I could see that science by itself did not have all the answers to the human condition, that ethical values and motivation had other roots, not contrary to science and reason, but complementary to them, creating a coherent whole. I then asked myself how could I capture my own journey to hope in a way that might make it accessible to others, particularly those of a rational, sceptical mindset? This book is the result. May it stimulate your thinking and open your heart to the better world that is possible and your own role in building it.

Introduction

So, you want to change the world?

Previous generations have made a mess of it. There is hypocrisy, injustice and corruption everywhere. Some people say that this is part of human nature and that it cannot be changed. Many are without hope and have given up. But you are not ready to abandon the world and withdraw into your shell. You want to make a difference. But how?

This is a do-it-yourself guide for the lost seeker, budding activist, or someone looking for greater meaning in life. It does not have all the answers; those, you will have to find for yourself. But it does try to assist you to ask the right questions and it has tools that might be helpful along your journey to finding those answers.

The journey will be hard. There are no quick fixes or easy solutions to the problems of the world or to the challenges of life. Changing the world into a better place will take centuries and you will not live to see the end of it. In any case, there is no 'end', only another starting point in the cycles that mark the ever-advancing journey of humanity. But, someone has to start; so why not you?

You are not alone. Thousands of others feel the same way as you and are starting along similar journeys. You will be sure to cross paths with some of them and will maybe even join up to travel a part of the way together. And however long the duration, whatever you accomplish on this journey will not be lost, because others will come after you and pick up wherever you left off. This is a voyage of hope and your life can be part of it.

It does not matter where you start

You may be at the point in your life where you are studying to prepare for a career, but are realizing that the professional world today is competitive, highly pressured, materialistic and stressful. It would seem that you should be able to advance on merit and ability, but there is a general lack of morality, and so lying, cheating and corruption too often have the upper hand. Or you may be at a turning point, dissatisfied with the life you are living and looking for a new direction that has more meaning.

At the same time, you have a moral yearning and desire for social justice and long to be an activist with a life of purpose, bringing real change in the world and overcoming the forces of negativity that surround you. Yet, it is difficult to go beyond what can too easily become a sterile intellectual debate and really do something when the problems seem so overwhelming. The resulting spiritual vacuum and inner pain can make you deeply unhappy, despite the opportunities around you.

Or, you may find yourself unemployed or unemployable, faced with a future worse than that of your parents. Perhaps war, violence, persecution or economic necessity have sent you on the road to exile and you are a refugee or migrant, ready to risk anything to try to find a better future. Regardless of where you are in your life, it is not easy to be starting out in the world of today.

Life is a journey

Life is a voyage through time and space. You are born, grow up and live and die in a linear progression of time, through various stages of growth and decline, of stability and of rapid or even sudden change.

Few people today spend their whole life in the same place, so you may move around your country, or even the planet, experiencing different geographic and cultural environments.

We seldom take the time to stand back and reflect on this whole process in its entirety. If you are young, you just want to 'get on with it'. You may not even ask how much control you have over your own life. But are you just being swept along with the current, doing what comes your way or the same as others are doing, or are you consciously choosing where you want to go and what you want to make of your life?

In a rapidly changing world, it may be impossible to know your ultimate destination and what the future will bring, but there is still a lot you can do to be able to say finally whether you have wasted your time or have had a life that was well-lived.

Life is a journey and, like any journey, there is a lot that you can do to prepare for it and to make the most of it. That is what this story is about. It is a user's brief guide to life and what it takes to place the maximum chances on your side so that your life will meet your highest expectations. It builds on the knowledge of science and some of the latest thinking about complex systems, but also examines fundamental questions of ethics and values about why you are here and what your fundamental purpose is as a human being. It places these individual challenges in the larger framework of a globalizing world that is experiencing multiple crises. It looks for paths to sustainability and a better future, exploring what you can do to go in that direction. It is, in a sense, a guide to how to bring greater meaning into your life.

The voyage of life is not easy. From the pain your mother suffered as you were born into the world to the inevitable decline of old age and ultimately death, the trek of your life will lead you through perilous valleys and up exalted mountains, with many opportunities to fall short or to lose your way. Some of the challenges are within you; others lie in wait along the way. At times, it may seem easiest to withdraw, to look for the easy way out, or to follow the crowd.

We are living in a challenging and often frustrating age, where hope is a rare commodity. Whether you are young and just beginning on this journey, or are already somewhere along the way, the obstacles may seem impossible to overcome. Yet, this universal quest for fulfilment, although arduous, can also be exhilarating, equipping you for the next challenges along the road and offering ever-growing vistas of new potentials as you forge ahead.

In telling this tale, the aim is not to lay out a wide highway for everyone to follow, like a herd of goats or a rush of commuters, but to provide some signposts to help you find your own unique path to your destiny. In particular, at a time when world crises seem to multiply and catastrophes seem imminent and doom inevitable, it is essential to put all these problems into perspective and to recognize that they are the symptoms of a time of transition. Rather then deny their reality, you will seek to understand their role and purpose in cleansing a sick society of its ailments, preparing for a new world civilization to emerge.

This is a story of courage, of the weapons of science and faith that are necessary to fight demons within and without, of environmental, social and economic valleys to cross and mountains to scale and of the peace, wisdom, justice and love that are the fruits of a life well lived.

Seven valleys

In keeping with past traditions, this traveller's narrative will take you across seven valleys and up seven mountain ranges, each addressing one of the critical dimensions of a world in crisis.[1]

First you will cross the desert-valley of blind souls – where individuals are crawling over each other for power and wealth – before scaling the mountains of long-term vision, integrated knowledge and evolving systems. Here you will acquire the shield of ethics and the values necessary to defend yourself along this journey.

Then you will pass through the second valley, with the threatening jungle of environmental crises. You will edge along the precipices of planetary boundaries, catastrophe threatening at every turn, before hiking up towards the peaks of environmental sustainability.

Beyond is the third valley, a social wasteland of the wealthy who live in castles besieged by the masses of the poor and disempowered. Here you will pass a dangerous crossing which will conduct you to the foothills of community building and up the path of social action to the mountains of justice and unity.

The strength acquired on this journey will equip you to face the fourth valley, which is filled with fire and smoke. Here the perilous quicksand of a materialist economy and a consumer society will threaten to suck you ever deeper into debt. But, beyond this valley rise the majestic peaks of the social economy, richly forested with work – work that is done in the spirit of service.

The experiences and trials of these valleys will help you to traverse the fifth valley, which is crisscrossed with canyons of self and ravines of corruption, so that you can begin to climb the cliffs of altruism and scale the heights of service.

As you continue in your search for greater reality, the sixth valley is filled with the mists of vain imaginings from which you must find your way towards the high plateau of inevitable human limitations. But it is here that you will glimpse a view of the unattainable universe of higher realities and infinite potentials.

From there you enter the seventh valley, wherein lie visions of the future that are within your reach, inspiring your steps ever upward on this earth while refining your character and contributing to an ever-advancing civilization. The higher you go, the greater the vistas of the emerging world civilization stretching out before you and beckoning to future generations.

While your individual journey will only contribute to a short span of social progress, along with billions of others your fulfilment will come from the qualities you have acquired during your quest and the bricks you will lay in the edifice of the new civilization yet to come.

Preparing for your journey

You can use this guide to your *Pursuit of Hope* in many different ways. You can read it like a novel, stopping to reflect from time to time, but looking forward to what may come next. Or you can read just a little bit each day and think it over. The many sections make reading easier, even on a smartphone or tablet. Some sections are longer and may be all that you can manage at one sitting. Short sections could be read a few at a time. If you read two sections a day, you will complete your journey in about a month. In a few plaes there is more text in boxes for those who want to delve deeper into a topic; or you can just skip the boxes and move on.

A word of warning is necessary. As with any voyage, you should not start out overburdened with excess baggage – which you will only have to abandon along the way. We all start out with inherited assumptions about the world and about our-selves, transmitted by our parents or unconsciously absorbed from our community and the media, perhaps also including a long-standing religious tradition or political perspective. In this journey, you need to be ready to question – not to reject eve-rything in a violent gesture of liberation – but to bring to the surface and re-examine each assumption and preconception in an act of independent investigation of truth, so that what you keep is your own and forms a coherent whole, upon which you can rely throughout your journey.

Equipped with a moral compass and signposts for this jour-ney, you can set out with confidence and hope, undeterred by the pessimists and nay-sayers or the fearful and jealous, empowered with the strength to take responsibility for your own destiny.

I

The Valley of Lost Souls and the Quest for a Unifying Vision and Values

This first valley will seem familiar because it resembles the world we live in. However, the people in this valley are blind, even though they have eyes; theirs is a blindness of the intellect and of the spirit. They are wandering in the desert, chasing mirages.

They see wonderful visions of material happiness. In order to be happy, they must buy the latest electronic gadget, or apparel that everyone must wear to be seen as part of the group, or music that affirms their identity. Yet, once in hand, the mirage vanishes; there is always an even newer gadget, or style, or song beckoning out there, offering even greater satisfaction and calling 'buy me, buy me, buy me'.

This valley traps most people as passive consumers of material goods. They keep the economy growing and feed ever-increasing wealth to the giant corporations that rule it. One cause of this blindness is that these people see only the material dimension of life and seek satisfaction only in hedonistic pleasures.

A second cause of blindness comes from tunnel vision – seeing only a small part of the whole, determined by one's field of knowledge, professional training or experience in life. An economist sees only economics and filters everything to fit into that field of vision. A political activist will see only what reinforces her or his ideology and sees everything else as a threat. A religious person may be immersed in his or her own tradition and belief and completely ignores the richness of spiritual discovery in other faiths or denominations. A scientist may be convinced that anything not validated by the tools of her or his discipline or published in a peer-reviewed journal is not worthy of attention. Such tunnels become traps that do not provide any way out of this valley.

The blindness of the intellect rejects science and knowledge when it is inconvenient. For those who are blind as such, truth is whatever you want it to be and has no relation to any objective reality. The ends – usually power or money – justify any means. This anti-intellectual blindness can be very calculating, using the tools of science for its own ends. It can also be very popular, because seeing everything in black and white, with the illusion of absolute certainty, brings comfort.

Blindness of the spirit shares many of these characteristics, but it denies any spiritual reality and sees humanity only as animals with physical needs and desires that can be satisfied with carnal pleasures and material wealth. It denies any higher human purpose or meaning and sees our existence only as the result of random processes.

Yet, science and history both show that in every human society there is a fundamental need and desire for meaning in life, a search for the self and the universe, with connections that give life a purpose. Spirituality, in this sense, refers to the natural and universal need to understand the world and our place in it. This spirituality can take many forms: secular or religious, in nature and the sciences, in art, music and poetry. When a person has no sense of meaning, or rejects this need for meaning, the vacuum in their existence leaves a fundamental anxiety that is too easily compensated by aggression, depression and addiction.[1] These are, in fact, the symptoms that define this desert of blind souls.

This blindness of the spirit is most evident in the rejection of anything that might fall within what has traditionally been called God or religion. Yet religion is the institutional response to the human need for spirituality; it is a social institution that meets a natural and universal human requirement.[2] This is not to ignore that many of the worst atrocities we can imagine have taken place in the name of religion. But has religion not been used in such cases as a mere cover for baser human motivations?

To overcome this blindness we need to look objectively at religion, not as something that is inevitably anti-scientific and anti-reason – as it is too often in the religious traditions of today – but as a potential force for good: for unity rather than division and in harmony with science and reason, while providing answers to questions of human purpose and motivation. It is within this framework that you will encounter religion and spirituality along this journey, freed from the trappings of the past and able to play a constructive role in transforming civilization. If you have an *a priori* prejudice against religion, try to set it aside for the duration of this journey. You can always return to it later.

Diagnosing the illness

It may help at the start of this journey to imagine that you are in space looking down on this valley, as if it were the planet Earth, with a vision that allows you to see every single human being and to understand what is in each mind and heart. You can look over all of human society and can seek to identify the ailments from which it suffers.

Just as a doctor examines the symptoms of a patient in order to diagnose an illness, it may help to gaze out at the vista of the seething masses of humanity around the world. In one direction there are bitter enemies engaged in warfare, with hapless victims desperately trying to migrate to safety. In another there is organized crime, with corrupt leaders seeking to extract as much wealth as they can from others, regardless of the harm caused. In yet another direction, the rich are becoming ever richer, while the young can no longer find work; the marginalized are ignored and even the middle classes are slowly sinking into poverty. From behind come the sounds of the moaning of prisoners of conscience, victims of injustice and those jailed for their beliefs or their efforts for freedom and human rights. You may ask why you see so much poverty everywhere when the world clearly has enough wealth to meet the needs of everyone. Or perhaps you ask why there is so much fear of strangers, immigrants and the 'other', when all people are members of the same human family.

Despite the physical unity brought about by technology, the world is still fragmented into nations, classes, so-called 'races', political parties, labour unions and employers' federations and all sorts of other groups. Most of the people who have grabbed power are short-sighted, selfish, greedy and materialist, often building their power using manipulation, fundamentalist or radical agendas, xenophobia and fear.

However, this vision before us is not totally grim and foreboding. Here and there in the valley we see glimmers of light in the darkness. Some individuals with pure hearts and strong consciences are resisting the outside pressures and living a life of virtue and goodness. A few neighbourhoods, villages and communities have the unity and solidarity necessary to educate their children, giving young people values to live for and the motivation to study to improve their understanding of material and spiritual reality. Aware of their higher or spiritual nature, they share prayers and devotions together and consult about practical solutions to their problems. Young entrepreneurs are starting social enterprises and even some large corporations are asking if there should be a purpose beyond profit. Some scientists are devoting their lives to improving our understanding of the world and the solutions to our physical problems. A few leaders really do have the welfare of every human being at heart. There are people who are selflessly giving to others in need. But, despite such actions, these efforts still fall far short of what is needed to change the present course of society. The fundamental illness remains. There is a general vacuum of ethical principles, human values and moral purpose.

Looking for tools

We need a conceptual framework in which to understand this complex system, within which an expanding humanity is trying to live on planet Earth. This will help us to ask the right questions. If science and technology have today eliminated the physical barriers between us and provided the knowledge to solve the age-old problems of scarcity and the struggle for existence, why do we still bring such horrors on ourselves? Is this our inevitable fate, or can we do things differently – should the desire and willingness, as well as the effort to change, be present? What is our human purpose?

We seem to have a much higher potential as human beings, as exemplified by some outstanding and exceptional individuals who are widely admired but seldom copied. Surely such behaviour would be too hard, would take too much of an effort? Yet, isn't such action a matter of free choice?

The voyage on which you are setting out will explore these questions and hopefully help you to find some answers. The metaphorical journey ahead is not just through some fantasy world, thought up for a computer game, but parallels what we experience in our own lives now and through all the years ahead.

In order to find your way out of this valley, you need two

sets of tools that will help you to succeed on this journey: systems thinking and some fundamental values.

In a world of increasing specialization and compartmentalization, few people are educated to think about the whole and the ways in which all the individual parts fit together and influence each other. We may think that everything works mechanically and predictably like a machine and that equilibrium is the ideal state, but the world and its components – including ourselves – are organic and undergo constant change and evolution. It is hard to deal with this complexity. Even computer programmes that follow mathematical rules can exhibit strange behaviours. Many of the crises in our society are the result of systems failures.

We usually look first to science and technology for solutions. Science can give us the means, but it does not have all the answers. The products of science can be used as easily for war as for peace, for domination or sharing, for torture or healing. Only a strong framework of values can tell us how to use the tools of science and what kind of a society we may want to build.

This dichotomy reflects our own reality. We have both a rational capacity for scientific thought and an emotional dimension with feelings and beliefs, of which the latter are sometimes far from rational. These two realities coexist both within us and within society. They can be antagonistic and in conflict, or, ideally, they learn to live in peaceful coexistence and can even build upon a larger whole. One of the first challenges of this voyage is to learn how to bring these two realities into coherence and to appreciate their complementarity. This is what will allow you to escape from this valley as you journey towards a better world.

Systems thinking

Perhaps the easiest way to explain a systems approach is to start with an example that we know intimately, even if we have always taken it for granted and not thought much about it: the human body.

A scientist might start by looking at physical properties, such as our weight, size, volume and density, or perhaps chemically at the amounts of elements like carbon, oxygen and nitrogen, or our molecular composition of proteins, carbohydrates, enzymes and water. Biologically we represent a collection of cells, tissues and organs, both our own and perhaps an equal number of cells of microbes and other organisms enveloped within our very skin and performing various functions like breathing and eating. Of course, we did not just appear out of thin air, but are a product of reproduction in the human species, with parents who produced a genetic recombination and gave us life, starting a complex cycle of growth, reproduction, ageing and death and in which we help to perpetuate the species from generation to generation.

We could also be viewed in terms of our behaviours and activities. A doctor would consider our state of health and determine the medical treatments required. We are not just a naked body, but may be distinguished by how we dress, how we surround ourselves with buildings and other structures and our ability to invent and use technologies that allow us to do things beyond our own physical capacities. For an educator, we started as a child in whom various skills were developed – like reading and writing – advancing to various spheres of knowledge and professional skills that allow us to contribute to society and to adapt to various conditions and situations. The psychologist would see in us a conscious being (whatever that means) with emotions, different kinds of intelligence, memories and a past history of experiences that influence us in various ways.

We also have what could be called a spiritual dimension, made up of our ethics and values, or perhaps a religious or cultural tradition, or other various sources of motivation.

As a social organism, we do not live in complete isolation.

We belong and contribute to many social structures, from our family and community to our nation and even our global society. Nor are we cut off from our surrounding environment. We take in food and drink, release waste and interact in multiple ways with both our natural environment and the city or community that many people over many years have built around us.

This brief summary of us as a complex system, dependent on the good functioning of all our internal components (themselves complex systems) and, in turn, an individual human being within many other complex social systems that exist at higher levels of organization, shows the challenges and opportunities of a systems approach. As we can see from this example, some of the elements of a systems approach include:

◆ Seeing things as dynamic and constantly changing, rather than static.
◆ Understanding the processes at work.
◆ Looking for causes and effects.
◆ Exploring interactions and understanding what influences something else within the system.
◆ Integrating all the different components in order to get a view of the whole.
◆ Looking for emergent properties that result from the whole system and which might not be predictable just by adding up the parts.

You will find that this systems approach will not only arm you to confront the dangers in the valleys to come on this journey, but will also help you to integrate what you learn in each valley into a deeper understanding of your life-voyage of discovery and achievement.

Information

Information in systems

This valley is full of something that cannot be seen until it is captured in a physical form, like writing or electronic bytes in a computer memory. This is information and we have more of it than ever before. We are living in what is sometimes called the information society, that is, a society where we are continuously exchanging, managing and using increasing quantities of information. It can be helpful to try to understand the role of information – intangible but fundamental – in the functioning of these complex systems.[3]

Each level of organization in our planetary system has its own ways of storing and using the information necessary for its structure and functions. For example, in chemical systems it is the structures of different atoms that determine how they combine and interact to build all the molecules of which matter is composed. Biological systems encode their information in the DNA of their genetic machinery, so that a single fertilized egg will contain all the information needed to generate the whole adult organism. In built or mechanical systems, engineers or architects have come up with a concept embodied in various plans and specifications which, if followed, will produce the same machine or structure each time.

By the time we come to human systems, the information on their organization is captured in statutes, laws, regulations, practices and customs, usually recorded in written texts or passed on in oral traditions. At the most fundamental level, human organization is guided by values, ethical principles, beliefs and cultures that provide the basic rules for society.

Communications

Looking at the dynamics of information and its transmission, we need to understand the different systems for the communication of information. Our body, for example, has multiple systems of communication: electrically through nerves; chemically through hormones and chemical signals; genetically with mechanisms to turn genes on and off; and probably in other ways beyond our present understanding.

There is also a danger of information overload. Thus, an effective system needs to be economical and able to communicate the information necessary. These systems allow higher levels of control and regulation, with feed-back mechanisms that indicate when to start or stop a process. One way to manage all of this is through nested sub-systems, with many autonomous functions within a system and with only the necessary exchange of information for cohesion between the levels. This is where the principle of subsidiarity is important, leaving as much control as possible at the lowest level closest to the needs and with only a limited amount of communications and authority at higher levels. Indicators are often used to signal the essential information content without extraneous detail.

System functioning and evolution

One important lesson from this perspective on information is its importance to system functioning and evolution. The higher the information content, the better the system works. Deficiencies in the information component can lead to catastrophic disfunction within the system as a whole.

For example, if intercellular communications fail in some part of your body and the cells there are left free to divide without limits, it will result in a cancer. Similarly, if laws are ineffective and basic ethical standards are lacking, the result is corruption and social chaos, something that may seem familiar from the diagnosis of society in this valley.

This shows the importance of values and spirituality as the most fundamental level of reality. If the basic rules of society are wrong or absent and each individual is driven largely by the selfish, animal side of human nature, the result is the social equivalent of cancers. Without change at this level, change at higher levels is difficult. Complex evolutionary systems thinking can give you the scientific foundation for understanding where you are and where you are going on this voyage through the valleys of life.

Solutions

An information systems perspective can also suggest some approaches to finding solutions. With better information flow and accounting, we can become conscious of the different dimensions of the systems we are in and how they work and can see more clearly where to intervene and how to improve them. New information technologies have fundamentally transformed the ways people can relate to each other and how they organize institutions and create knowledge. This is a potential that we are only just beginning to appreciate and that will unfold in the decades and generations ahead.

At the same time, we should not let the superficial attraction of today's information and communications technologies – manipulated for commercial gain and often designed to create dependence and addiction – hide the importance of change and exchange at other levels. Nothing can or should completely replace direct human interaction. While ideas are easily transmitted over the Internet, it is not always the case for emotions and life cannot be reduced to a few emoticons. Non-verbal communications, through gestures and facial expressions, express many things that words cannot. Even our odours signal immune system compatibility between potential mates. Robots and 'assistants' can never completely replace people.

Social interaction takes place largely through information exchange. While a mother and child start with direct physical contact, this moves quickly to verbal communications and eventually language. This is how we build and maintain family ties and the social relationships underlying communities. For most of human history this communication was face-to-face and in the last few generations by written letters. Now new technologies provide new possibilities, but also risks. Building families and communities is fundamental to our future and you will see other dimensions of this in the valleys to come. It is just necessary now to understand the potentials and limitations of information technologies so that they can become an essential complement to modern social organization at multiple scales.

Systems dynamics

Systems science is also shedding new light on how complex systems work and you will see examples all along your journey. Complexity can both increase resilience – the ability of a system to maintain or restore its balance – and vulnerability – which can lead to system collapse. We need to understand how to increase the former and reduce the latter.

A system that has lost its natural controls may be subject to what is called overshoot and collapse, where a rapidly growing population may suddenly exhaust resources fundamental to its survival or destroy basic elements of its life support system and then die off. Systems often progress through what are called punctuated equilibria, meaning that a system can seem relatively stable and gradually improve its efficiency adapted to its environment at a certain level of organization. Changing conditions or a shock can lead to a period of rapid and perhaps chaotic change until new potentials allow a transition to a new period of stability. In the age of the dinosaurs, reptiles evolved to become the dominant life forms adapted to their environment, until sudden change – probably including an asteroid strike and extensive volcanic eruptions contaminating the atmosphere – caused their extinction and allowed the rapid evolution of birds and mammals.

The same thing applies at other levels of organization. The cells in our body age, develop faults and are replaced. We all ultimately age, our bodies will wear out and we shall die – but hopefully after most of us have produced offspring to replace us and to maintain the continuity and progress of our species. Sustainability in human society comes through reproduction and education, transmitting our collective knowledge from generation to generation. Effective systems go through a process of self-renewal, even if their individual components are constantly turning over.

There seems to be a natural evolutionary progression, with simple systems with few components and little information evolving into more complex and efficient systems with a larger information content. Innovation, whether through genetic mutations or human invention, makes this progression possible. This suggests that we should not be seeking security, stability and comfort in our society, but that progress comes through a culture of change in a learning community, a theme that you will come across again in subsequent valleys. At the present time in this valley, when change is accelerating, you will need to continually adjust and adapt to new situations and opportunities, rather than clinging to past certainties that are no longer relevant.

Another important lesson is the importance of reliability or trust. If a system can be counted on to perform reliably at its level, very little communication is needed with other levels and the system is highly efficient. Translated to a societal level, if people can be trusted to do the right things, very little control is necessary. Bureaucracy evolves in unreliable situations to maintain control, becoming increasingly inefficient as people become untrustworthy. The concept of trust shows another systems characteristic: trust takes a long time to build and is all too easily lost. Complexity takes a long time to build and evolve, while it is much easier to cause its unraveling, decline and collapse.

The most important lesson to retain at this point is the importance of systems thinking. This is a whole new way of looking at problems and their solutions. You will find that this new perspective will often be useful on your journey as you face the challenges ahead.

Values

If you like where you are in this valley and decide just to stand still or sit down, remaining comfortable in your blindness, you will never leave and discover the challenges and opportunities that lie beyond.

One of the things that emerges from a systems perspective is that the processes are more important than the content. For example, crudely put, it is not how much money you have that is important, but rather how you use this money. And behind the processes are the rules by which they operate and the control mechanisms that determine what happens when and by how much. The more systems are self-regulating, with internal controls, the less they need controls imposed from outside.

Self-regulation

As scientific understanding of the genetic code behind all life grows, it has become clear that, in fact, only a small part codes for the structure of specific proteins and enzymes that are the building blocks of living things. Most of the DNA determines control mechanisms for when particular genes should be active; in other words, where and when each block should go into the building. It is the difference between the mason simply laying each block in turn and the architect who has the plans for the whole building and gives instructions as to what to do next. We are still far from understanding these genetic mechanisms of self-regulation.

Looking at human systems, what forms do the rules of systems organization take? In a complex society, many of the rules are codified in laws that determine legitimate forms of behaviour: how we support the government through taxes, how our employer can treat us and what to do when two cars meet at an intersection. There are, for example, rules for traffic to prevent people from causing accidents and killing each other.

More fundamentally, it is the values, ethical principles and moral codes that are the basic framework for social organization. Understanding what the values are and how they are expressed in the system is essential. Without this knowledge, we may deal with the symptoms of social problems without getting to the root causes.

Where do our values come from?

In most stable societies values are transmitted from one generation to the next, first within the family, then by social institutions like schools and religious education and, today, increasingly via the media. Small children first learn by observation and example. What parents do is often more important than what they say. Later in a child's development peer pressure becomes important. We want to belong to a group and be accepted, so we adopt the values of the group. Pre-adolescence is a critical time in the formation of values; our parents become less important and we begin to choose the values that will determine our adulthood. Up to this time, values are assimilated unconsciously and this can continue as adults, with values taken for granted because 'that is the way things are'. It takes a great effort to change values once they are set in this way, which is why it is so much better to start with positive values to begin with.

One worrying phenomenon in many countries is the breakdown in the transmission of values within the family. Many of today's parents, at least in Western secular societies, rejected religion in their youth and thus failed to transmit any knowledge of the values derived from religion to their children. Yet this is the domain that speaks about the meaning and purpose of life. Children are thus left adrift, trying to fill the vacuum of values and purpose from any source – from drugs to fads to fanatical movements. Another consequence is that ignorance of religion leads to fear and rejection, especially of those we deem to be the 'other', such as immigrants and people with different beliefs from the mainstream. Some school systems have instituted compulsory religious education in the primary years, objectively presenting all the major religions to reduce this source of social tension.

This life voyage you are embarking on will therefore primarily be an exploration of values and we will use metaphors throughout the story to make this invisible dimension of your life and society more visible. Only a part of social organization is written down in laws and regulations; much more is imbedded in culture, received wisdom, verbal tradition and example.

What are values?

Values can be defined as qualities upon which worth, desirability or utility depend. They are principles or rules generated by an ethical or spiritual framework. Values are what determine how humans relate to each other. They are the social equivalent of DNA, encoding the information through which society is structured. Any change in society has to start with a transformation in its values.

A value or perception that is maladapted to the needs of society can be an obstacle and cause great damage to the social fabric. For example, a racial prejudice that labels all people of a particular skin colour as inferior, untrustworthy or even dangerous, prevents any constructive interaction and marginalizes the excluded group, sometimes even becoming self-fulfilling in the reaction it generates in others.

Sometimes a reaction that may have been justified in our ancient past needs to be mastered and redirected within the complex societies of today. The fear reaction of 'fight or flight' makes sense when face to face with a wolf, bear or lion, but not when triggered by an irrational fear of someone else who is just different from us. Research suggests that people who are more prone to such fears and react more acutely are also more defensive ideologically, conservative politically and unwilling to question their assumptions and consider new ideas and perspectives. Learning to deal with this fear is a prerequisite for change at other levels. In this voyage, you will be confronted with many kinds of fears which you need to overcome in order to advance.

These values, whether constructive or destructive, in terms of social evolution, operate at two levels: the collective and the individual. Collectively they are expressed not only in laws but in social behaviours and the organization of communities, both consciously and unconsciously. For example, racial segregation can be legally mandated, as it was under apartheid, or can simply occur because particular groups cluster by neighbourhood and others move out to avoid them. This may just be an organic expression of widely held beliefs, but the consequences of ignorance, fear and intolerance are socially very destructive.

Changing the structures of society requires an evolution in these beliefs. This can be more difficult when social contact is reduced. How can you discover that the 'other' is just a person like you if you have no chance to get to know them? Even more tragically, political manipulation based on fear of the 'other' can tear apart communities where people happily lived together for generations, leading to conflict and deep social scars that can take generations to heal.

Individually, it is your values that determine how comfortable you are with others who are different from you in culture or physical characteristics. The stronger your framework of positive values, the more resilient you can be in challenging circumstances. Discovering what your values really are at the deepest level is a first step and an ongoing challenge as you progress, for there are always surprises hidden in your unconscious. The fact that these values are often unconscious, 'because that's the way things are', makes them particularly hard to confront and change. People are very good at self-deception and are easily influenced by others and you must become more mindful of your reality.

The battles you will need to fight on this journey will often be at the level of values and you will need the weapons of positive values. These battles may at times be as much with your own self as with others you encounter on the way. This will be a recurring theme through all the valleys, as it is fundamental to reaching your ultimate destination.

Ethics

The key to progressing in the subsequent valleys on this journey will be the strength of your personal ethical code or framework of values. There will be many traps and pitfalls along the way and temptations to take the easy way out or to just give up. You will need strong resolve and force of character to overcome the obstacles ahead and deeply held values can provide these. Your journey will be an opportunity for you to strengthen your own set of values that you can then carry with you throughout your life.

Every society has some ethical framework in which its values are expressed, as good or bad and right or wrong. Without this, society would descend into anarchy and disintegrate. The theory of group selection suggests that values like altruism, which might seem contrary to self-interest and reproductive success in purely evolutionary terms, survive because they strengthen the group – and individuals in a strong group have a competitive advantage over isolated individuals or those in a weaker group.[4] In the past, when tribal groups or nations competed at a frontier between them, social cohesion was selected for as it provided the strength to resist the enemy.[5]

Moral values

Moral values state what is regarded as being good and of primary importance to human civilization. They are often articulated as ideals and define what is right from wrong. A capability of moral reasoning starts from abstract general ethical principles to resolve conflicts that arise from moral dilemmas and ethical problems. Ethical principles are therefore the operational expression of moral values and provide guidance to decision making and action. Ideally, such principles should be general and not limited to specific individuals or associations. They should be universal in application and publicly known and accepted. They should impose an ordering on conflicting demands and have a condition of finality, in which they become the final court of appeal in practical reasoning.[6] There are many sources of ethical principles: philosophies, individual belief systems, indigenous or traditional cultures, major religions and alternative belief systems such as nature worship, deep ecology, new age, etc.

A central ethical principle in successful social organization is justice. Justice is the first virtue of social institutions, as truth is the primary virtue of systems of thought. A theory, however elegant and economical, must be rejected or revised if it is untrue. Likewise, laws and institutions, no matter how efficient and well-arranged, must be reformed or abolished if they are unjust. The right to justice should not be subject to political bargaining or to the calculations of social interest. As the first virtues of human activities, truth and justice are uncompromising.[7]

Values are put into practice in society through mechanisms of reward and punishment. Good behaviour is rewarded by social acceptance and breaking the rules leads to punishment through a system we call justice. Both law and ethics are concerned with the application of justice. Law is based on explicit legal texts, courts and institutions for enforcement. It is a top-down regulation of society, based on punishment and the use of force, if necessary. It is a costly way to maintain society. Ethics is founded in the individual attachment to the principle of justice and its application. It is bottom-up and self-motivated, relying on reward for good behaviour more than punishment. The stronger the ethical framework and its application, the less the need for law. Strengthening the ethical framework in a society is a more cost-effective, process-based solution.

Just as a computer programme determines how data is processed and results are generated in a computer, so can an ethical framework be considered the social programme with the rules for the proper functioning of society. How then might we 'programme' human beings to be effective contributors to an evolving civilization, with increasing levels of cooperation and reciprocity, more social cohesion and higher efficiency? You will discover answers to this question all along your journey.

Ethical failures

In today's society, with globalization as a force uniting the world's peoples and rampant materialism breaking down traditional ethical structures, the world is suffering from multiple ethical failures. Without an ethical or moral conscience the negative forces in society become more powerful. Centuries ago, Machiavelli already described a society based on the drive for power, expansion and material prosperity. We see too often today that the ends justify any means, with people or groups seeking power by force, violence, terrorism and war. There are endless scandals involving corruption, fraud and tax evasion. There are entire illegal economies that exist in parallel to the official economies that are captured in government statistics.

Part of the problem is that there is no inherent framework of values for institutions equivalent to the individual conscience. Governments are untrustworthy and often do not respect their obligations, except when they are forced to honour their debts – no matter how unjustly they were acquired. Businesses pay lawyers and accountants to find ways around the laws. In our materialistic, consumer-based society, the consumer is manipulated and ideally driven to addiction to tobacco, alcohol and junk food and even to computer games and social networks. Business leaders increase their bonuses while laying off workers in the interest of profit and 'productivity'.

There is also a polarization in many countries between opposing sets of values, leading to conflict and confrontation, as if values are by nature incompatible. In fact, they are generally complementary and need to be balanced. One area where the trap of ideologies and self-reinforcing communities of interest is most obvious is in politics. Almost everywhere with a certain level of political freedom there is a split into right and left camps, one emphasising the importance of individual freedom, responsibility and initiative and the other focused on social justice and action to remove barriers to individual advancement. In fact, these two poles of a single spectrum are complementary. A healthy society needs individual freedom and initiative, as well as collective solidarity. In a world of technological advancement generating ever-increasing wealth, there is no need to fight for a share of the pie – since everyone can contribute to a bigger pie – but rather to ensure that everyone has an equitable share.

Changing values

A society that is unethical is not necessarily without values. It is just that the 'higher' values that are necessary for a society to advance in cohesion, justice and material and spiritual wellbeing have been replaced by values of self-interest and the struggle for existence that are rooted in our animal nature. This raises the question of how to change socially dysfunctional values into those that lead to social advancement. How do we challenge assumptions about human purpose and goals in life and offer something more inspiring and motivating?

First we need to identify those values that contribute to the betterment of society. Truth and justice are a good place to start. Truthfulness and related values like honesty and trustworthiness, are essential to healthy social relationships and economic success. Every contract depends on being truthful and on those signing it being trustworthy. Justice is necessary to motivate people to work together. It finds expression in related values like generosity, solidarity, compassion and equity.

It is not sufficient just to learn about values. We need to see their importance and integrate them into our own beings by putting them into action. It helps to have the support of a community that shares these values, protecting and nurturing us in acts of service and accompanying us as our values grow within us. This can be a circle of friends, a faith community or an association, among others. As we grow from the knowledge of desirable values into belief in their power and efficacy, based on personal experience, we can acquire the wonderful combination of conviction and aspiration that can lead to committed and sustained action. These values can empower us to be active agents of our own learning through service to others.

Changing values is not easy. Science has begun to address how the brain works to resist change, even in the face of strong evidence that change is needed. For example, scientific information about climate change does not motivate action. In fact, the magnitude of the threat produces the opposite – an increasing denial of the facts and a refusal to act.

On the social side, we follow the thinking of the in-group we belong to, listen to people we trust, risk rejection if we try to think differently and thus often fail to see the whole picture. We may see the issue as remote and a conspiracy that threatens our way of life. Our brain responds poorly to distant, diffuse or uncertain threats and falls back on short-term certainty.

Our confirmation bias means that we select the information that conforms to our existing views and deny the real source of our problems.[8] What is even more difficult is to challenge values that are held subconsciously. We may not be consciously aware of some things that are really important to us and that we simply take for granted. We may even hold strongly to values that are in conflict with one another, without being aware of this.

A first step may simply be to try to bring your values to the surface and to recognize their importance to you. You may need to challenge your assumptions and reflect on the alternatives. Recent research has identified indicators of values that can be used to make them visible and to show how they are put into action in social relationships[9] and shared through education.[10]

With so much chaos linked to the lack of common values, is

there an alternative? And what might this look like? The valleys you will cross on your journey will help you to explore some solutions. Armed with the new tools of systems thinking and aware that you need to think deeply about your values as you pursue your journey, you are ready to leave behind the desert of blind souls and to climb up to the valley above, ready to face a new set of challenges.

2

The Valley of Environmental Crises and the Plateau of Sustainability

The second valley is the valley of environmental crises. Leading to the high plateau of sustainability, it is densely carpeted with jungles, dangerous swamps and polluted lakes of many colours. It is ringed by steaming volcanoes that may erupt at any time and swept by terrible cyclones powered by global warming that flatten everything in their path. Anyone attempting to leave this valley must cross chasms and precipices of planetary boundaries, with the risk of falling over the edge at any moment. Crossing such a valley carries great risks.

Instead of traversing this valley, you may find yourself dreaming of escaping from this planet and going to other worlds. Science fiction is full of such stories, often to help us better understand our own human reality on this planet by comparing it with imagined life on other planets. While technology has now made the vision of space travel a technical reality, it is also clear that the costs and risks, as well as the inevitable human limitations of time (a lifetime) and speed (much less than the speed of light), mean that only a handful of people could ever realistically go beyond our biosphere. Humans may be able to colonize other planets, but we cannot escape the one we are living on.

This means that, for the foreseeable future, we are stuck on planet Earth and, together with all the other humans here, will have to muddle through with what we have. Your life quest will inevitably take place on this Earth, within our planetary boundaries.

The environment therefore places limits on how we can grow and flourish and the more people there are on this planet the more we have to learn to share and consider others as well as ourselves. In this valley we all have to join the collective effort to confront these environmental constraints and find ways to overcome them.

In most cases, science has already told us what needs to be done. It is not possible to exhaust all the possibilities – as human imagination and creativity are without limit and new discoveries cannot be anticipated – so we can, at least, reassure ourselves that there are solutions out there. The impediments are not scientific but human and therefore not beyond human possibilities. At the same time, you can play your part by adopting a responsible and sustainable lifestyle for yourself and encouraging others around you to do the same.

Overshoot and collapse

The threats in this valley are largely of our own making. Environmental problems may seem like a recent development for humanity, or at least they used to be largely invisible relative to other priorities for human survival. Their emergence is due in large part to the extremely rapid expansion of the human population over the last two centuries and technological developments that have accelerated our consumption of resources and our impacts on the planet – which have now surpassed natural processes in their significance.

We have hit planetary limits, with the potential for what is called overshoot and collapse. In population ecology, overshoot and collapse is a well-studied phenomenon: a population that escapes from natural controls with regard to its numbers continues to multiply until it destroys its food supply, or some other vital resource, at which point its numbers collapse or it may disappear entirely. This can happen when a species is introduced into or invades a new ecosystem without the necessary enemies and parasites that provide natural population controls.

There is no biological reason why humans should not be vulnerable to the same phenomenon. The discoveries of science have reduced traditional controls – like epidemics and famines – on human population size. Ever since Thomas Malthus and his *Essay on the Principle of Population*, written in 1798, there have been predictions of the exhaustion of resources and the collapse of the human population.

Limits to growth

In the early years of the environmental movement, the report to the Club of Rome on *The Limits to Growth*[1] used computer systems modelling of population and economic growth and resource consumption to prepare various scenarios of the future. One scenario traced 'business as usual' – with population and economic growth continuing indefinitely; another showed that limiting economic growth, resource consumption and pollution would allow transitioning towards sustainability.

Economists have long refused to acknowledge that any closed system, such as planet Earth, has limits, or to recognize that our population, industrial production, food consumption and waste generation have to stay within those limits in order to be sustainable. They assume that market mechanisms and technological innovation can solve any problem. However, the model put forward in the report suggested that business as usual would lead to the collapse of both the economy and human population by the mid-21st century, thus requiring limits to growth. Since the idea of limiting growth was anathema to the economic and political orthodoxy (and still is for many), the report was attacked, derided, distorted and discredited (much as climate change has been attacked by sceptics today). The warnings were ignored and governments proceeded with business as usual, focusing on economic growth and ignoring planetary limits.

Technological innovation did have an unanticipated impact in the green revolution of the 1970s and 1980s, which increased food production, particularly in Asia, with improved varieties and chemical inputs, so the predicted famines have not (yet) materialized. However, such intensive agriculture gradually destroys arable soil, depletes nutrients, organic matter and the soil microbiome and accelerates soil erosion, leading to over a third of agricultural land becoming degraded in half a century.[2]

The same team updated their model and predictions at the time of the Rio Earth Summit in 1992, calling their book *Beyond the Limits*.[3] They estimated that society was overshooting sustainable planetary limits and living off its capital, but that time lags in the response of natural systems meant that the effects would not be apparent for some decades. A further *30-year Update*[4] maintained the same conclusions, that civilization would face major disruption in this century. In 2012, a comparison of their scenario projections, together with actual data from over 40 years, showed that their projections were right on target. The only difference is that today a smooth transition to sustainability is no longer a realistic possibility.[5] With too many people and too few resources under the current system to meet everyone's needs, a collapse of some kind can no longer be avoided.

The collapse of civilization

Other well-informed scientists have explored more generally the instability and uncertain future of Western civilization.[6] Making comparisons with the decline and fall of civilizations in the past, they often draw disturbing parallels with our present situation and question the common assumption that our science and technology will find a solution for every problem. An historical analysis shows the cycle of the rise and fall of societies as they build social cohesion under threat and then lose it again when things seem to be going well,[7] predicting that the tipping point of our own society into crisis and collapse could come by 2020.[8]

War

The acknowledgement of environmental limits is required as you face this valley, reflected in the enormous virtual mountain of dead bodies slain over centuries in the struggle for limited environmental resources. War has been with us since long before history started to be written and most history is about wars. Wars always have complex causes, but one of the underlying drivers is the tendency of any species, including us, to multiply until we reach the limit of our food supply and other resource production potential. The main difference between humans and animal populations is that, rather than waiting to starve, resource-short humans go looking for other humans to pillage and slaughter.[9]

In more modern times, leaders have engaged in land grabbing long before hunger has struck their population, as power and wealth increase the desire for more power and wealth. Since land and its resources represent wealth and people (particularly soldiers) have been regarded as expendable and replaceable, wars have continued. Only today has science changed the paradigm irreversibly, with weapons of mass destruction making large-scale war impractical, even for the victors, if not ethically unthinkable. The small-scale wars that continue today are almost always for areas rich in resources (usually oil, ores or gems), but, for the moment, greed has replaced hunger as the main driver.

The way out

Since your quest is one of hope, of replacing negative ways with something better, you need to look at the alternatives to overshoot and collapse at a planetary scale. Research groups that prepare global scenarios for the United Nations and other organizations have also explored alternative possible futures across the political spectrum, from 'business as usual' – with no economic or social changes – to countries closing their borders and retreating to some kind of fortress society that keeps others out, to a collective transition towards a sustainable future.[10]

These scenarios tend to show the unsustainability and vulnerability of our present forms of development, but they also demonstrate that a transition that avoids the worst of the problems is technically possible. Some take a positive view that the necessary transition is an opportunity to build a more sustainable civilization,[11] and whole groups of researchers, such as the Great Transitions Initiative, have been working on how this could be done.[12] We are leaving our collective adolescence and it is time to find more mature solutions to resource limits than conflict. This will be one of the challenges in the next valley on your journey.

Global environmental and economic systems

Having emerged from the lower valley with its threats of catastrophe, you will need to traverse the deep forest of conflicting and interacting environmental problems without falling into the swamps of economic thinking – that deceptively look like solid land, until you step into them and begin sinking. This forest is both dense and dark.

Part of the challenge is in understanding the behaviour of such complex systems as human civilization and the planetary environment. Experts in economics, political science, information technology and industry tend to see probable solutions in their respective fields and thus feel no need to worry. However, biologists observing the rapid decline of natural systems tend to be more concerned. Biological systems can evolve and adapt over many centuries, but not at a time scale of interest to our own civilization. Ecosystems can collapse just as can human societies.

It is the specialists on complex systems, who explore the interactions between many disparate factors, who seem the most alarmed as they study the relationships of energy cost, climate change, water and food shortages, biodiversity loss, pollution impacts, the declining reserves of critical minerals and other constraints that face a population that continues to rise rapidly with economic success in some parts of the world. Studies on the possible collapse of civilization highlight the ease with which complex interlinked networks (like electricity transmission) can shut down and show the vulnerability of human society to events such as a virulent flu pandemic that could kill key technical personnel around the world.[13]

It is these complex interactions between economic globalization, climate change, energy and resource depletion, persistent poverty, social imbalances and human security that make it so hard for you to navigate your way through the forest. The United Kingdom's Chief Scientific Advisor from 2008 to 2013, John Beddington, warned in 2009 that the world will face a 'perfect storm' of problems in 2030, as food, energy and water shortages will interact with climate change to produce public unrest, cross-border conflicts and mass migrations.[14]

Each problem interacts with the others in complex ways. Partial solutions will not solve these combined challenges that threaten future sustainability. This makes action very difficult, if we are to avoid the risk of ecological overshoot, as it requires a fundamental transformation of human society.[15]

Environmental and resource problems are, of course, intimately linked to the economic system. Those resources that are traded in the world market, like food and fossil fuels, are subject to destabilizing price fluctuations, often due to speculation. On the other hand, the failure of the market to reflect the true costs of economic activities to the environment and human society is what drives much environmental damage. Climate change has been called the greatest market failure in history.[16]

As the economic implications of the costs of environmental damage become more clear, it is not easy to see where the enormous requirements for investment capital and development assistance will come from. One journalist summed up the challenge very well:

> On current trends . . . humanity will need twice as much energy as it uses today within 35 years . . . Produce too little energy, say the economists and there will be price hikes and a financial crash unlike any the world has ever known, with possible resource wars, depression and famine. Produce the wrong sort of energy, say the climate scientists and we will have more droughts, floods, rising seas and worldwide economic disaster with runaway global warming.[17]

This theme was developed further by the international governing body of the Bahá'í community, which wrote:

It is evident that the current defective world order has failed to protect the environment from ruinous damage. The deepening environmental crisis is driven by a system that condones the pillage of natural resources to satisfy an insatiable thirst for more.[18]

The problems scientists identify are worsened by the weaknesses in the political systems, internationally and nationally, which are incapable of managing such complex global problems and the unmanaged globalized economy, where the environment is not a priority and speculation and corruption are out of control. Improving global environmental and economic governance and our ability to manage the Earth's resources in a planetary context are thus a key part of any solution.[19] Preventing overshoot and collapse is the challenge of our generation.

Planetary boundaries

Emerging from the dense forest of environmental complexity, the way out of this valley is crisscrossed with the deep chasms and steep cliffs of planetary limitations. Our human population and civilization have rapidly grown to the point where they are reaching and even surpassing the environmental limits of our planet, or what have been called planetary boundaries, with great threats to our future. You risk being pushed over the edge at any time and must work to keep your footing and balance.

What then are the major planetary boundaries and what do we all have to do to stay within them? First, we have to acknowledge that both the biosphere that sustains life on this planet and our global civilization are complex systems, with each part linked to the others, so that no one boundary can be considered in isolation. Scientists have defined nine planetary boundaries and, for three – climate change, biodiversity loss and the nitrogen cycle – we have already overshot the boundaries.[20]

Climate change

The first ridge you venture on is the immediate and threatening challenge of climate change, driven by the human release of greenhouse gases, especially carbon dioxide, into the atmosphere. At this stage of your quest, you are facing a fiery ordeal of heat and cold, tempestuous winds, floods and drought.

Since we discovered, a century and a half ago, that fossil fuels were a cheap source of ancient solar energy to replace human and animal labour and build a high-technology civilization, we have been benefiting from this cheap energy subsidy, often using it wastefully. More recently we have realized that returning all that carbon to the atmosphere is rapidly warming the planet, changing the climate, raising sea levels and threatening many parts of the world with droughts and disasters.

The science of climate change

The science of climate change is quite simple. The biosphere – the thin layer of air, water and soil on the planet's surface that harbours all life – is maintained by a complex set of delicately balanced systems, which are still poorly understood. The atmospheric conditions that permit life to exist were themselves created in part by the action of living things. The early microbes and plants removed carbon dioxide from the atmosphere and added oxygen, making animal life possible. Dead plants, the remains of marine plankton and terrestrial vegetation were buried and fossilized as coal, oil, gas and their carbonate skeletons became layers of limestone, locking a significant part of the Earth's carbon away in geological formations.

Carbon cycles through the biosphere as plants take up carbon dioxide to make organic matter and animals and decomposers return the carbon dioxide to the oceans and atmosphere. Natural transfers in and out of the lithosphere (geological formations) are too slow to

---

have much impact on the carbon cycle. The balance between these processes has been upset by the extraction and combustion of fossil fuels (coal, oil and gas), returning large amounts of carbon to the atmosphere and to oceans that had long been out of circulation.

The significance of this for the climate is that carbon dioxide, along with methane (also containing carbon), is among the most important greenhouse gases and traps heat in the atmosphere in the same way as the glass in a greenhouse lets in light but prevents heat from escaping. The climate has changed in past geological epochs – with both ice ages and much warmer periods associated with rises and falls in plant cover and carbon dioxide levels – due in part to the Earth's orientation to the sun and to the changing positions of the continents, which affect the way the linked ocean-atmosphere system redistributes heat around the world. Research has shown that, even with these natural cycles, there can be quite abrupt changes between warm and cold periods. Now, our massive use of fossil fuels is raising the concentration of carbon dioxide in the atmosphere to levels not seen in millions of years.

While governments signed the UN Framework Convention on Climate Change at the Rio Earth Summit in 1992 and adopted the Kyoto Protocol to start limiting the release of greenhouse gases, the political will to make changes was not there and the release of such gases continued to accelerate. The failure to achieve more than a vague agreement at the UN Climate Change Conference in Copenhagen in December 2009 is only one illustration.[21]

Finally, as the evidence of human-induced climate change became overwhelming, governments adopted the Paris Agreement in 2015 and committed to reducing their greenhouse gas emissions. But their commitments were far from enough to prevent rapid warming. Most countries are far behind in meeting their commitments, with some even abandoning them as too damaging to their economies. Meanwhile, the climate is changing even faster than scientists have predicted, with rapid warming melting ice in polar regions. There are signs that we may soon reach tipping points where it will be impossible to go back. Despite the overwhelming scientific evidence, powerful vested interests – such as in the fossil fuel industry – are blocking much of the necessary action.

We are now so addicted to fossil fuels and the vested interests in maintaining the present economic systems are so powerful, that we are not responding with the necessary urgency. To illustrate the scope of the challenge, the carbon in fossil fuels that has already been discovered and that is ready to be exploited (not counting unconventional fuels like tar sands and shale gas) is five times the remaining capacity of the atmosphere to absorb carbon without disastrous climate change.[22] Our only hope is to leave 80 per cent of that carbon in the ground, but that is presently unthinkable for economists, politicians and oil-producing countries.

Even with the Paris Agreement adopted in 2015, governments have put off any binding international action on climate change until 2020. The science shows that, after a reasonably stable temperature for the last thousand years, the average planetary temperature has started to rise rapidly and projections suggest that we could easily overshoot the +1.5°C increase that scientists believe will mean major climate change impacts on the planet and human society.[23]

This represents an important threat to human security. The International Institute for Strategic Studies has said that if climate change goes unchecked its effects will be catastrophic 'on the level of nuclear war, with countries seeing falls in available resources and economic vitality, increased stress on their armed forces, greater instability in regions of strategic import, increases in ethnic rivalries and a widening gap between the rich and poor'.[24] World-famous naturalist, David Attenborough, in his official address to the UN Climate Change Conference (COP24) in Poland in 2018, said: 'Right now we are facing a manmade disaster of global scale, our greatest threat in thousands of years: climate change. If we don't take action, the collapse of our civilisations and the extinction of much of the natural world is on the horizon.'[25]

Climate change on this scale is already having significant human impact. There is increasing damage from extreme weather events, such as floods, droughts and cyclones, with less winter snowfall and water shortages in summer (resulting in more extreme wildfires). Conditions for agriculture and forestry will change significantly, with high costs of adaptation and fish stocks already under heavy fishing pressure may shift location or collapse. With rising sea levels now predicted to reach a metre or more by 2100, there will be flooding of low-lying areas and islands. Such impacts will result in the migration of millions of environmentally displaced persons (some estimates put the figure at 200–500 million), a scale of human displacement that will dwarf anything previously experienced, forcing countries to lower immigration barriers and accept the free movement of people as part of globalization. The Stern Report to the United Kingdom government estimated the annual cost of uncontrolled climate change at more than US$660 billion, or 5 to 20 per cent of global GDP, as compared to 1 per cent of global GDP for control measures for greenhouse gases.[26]

Unlike other global environmental problems, such as stratospheric ozone depletion –where the number of actors was limited and international agreement on control measures for the chemicals responsible was possible – climate change threatens the very basis of the global economic system, which was founded on the energy subsidy from cheap fossil fuels. However, it is not that solutions are not available. We could replace all fossil fuel use in 20 to 40 years with existing technologies for renewable sources of energy, not counting nuclear energy or biofuels that compete with food crops,[27] creating many new jobs in the process.

So much climate change is already happening that we cannot avoid it, but we still need to mitigate the driving forces releasing greenhouse gases, while we also adapt to the changes under way. The challenge is that this is a global problem that can only be resolved through concerted global action, but our international institutions are still trapped in a paradigm of national sovereignty and consensus decision-making, with each country defending its national interest rather than the good of the whole.

Biodiversity

If you make it across the ridge of climate change, you will face the slippery slope of biodiversity loss, with thousands of species already tumbling by you on their way to extinction. Our planet has already crossed this boundary and you risk being swept away as well.

Biodiversity loss

Life has been evolving on this planet for billions of years; first as simple microbes, then more complex plants. Then, once the plants added oxygen to the atmosphere, animal life became possible. Finally, humans emerged as a species capable of abstract reasoning and higher states of consciousness. Over time, many species become extinct and new ones evolve, with a natural tendency to increasing diversity and complexity. However, various things can interfere with this progression. There have been, in the past, a few mass extinctions with a large percentage of species disappearing, followed by a new evolutionary explosion as species evolve over millions of years to fill all the vacant places even better than before.[28] Some of these mass extinctions are associated with planetary catastrophes, such as massive volcanic eruptions poisoning the atmosphere, or the asteroid strike that probably wiped out the dinosaurs.

Today we are experiencing a new mass extinction, but this time resulting from human activities. As the human population has multiplied and our technologies have improved, we have converted more natural areas to other uses, spread invasive species and over-exploited species and other natural resources. This has raised the extinction rate to a thousand times the natural rate of species loss. Now with accelerating climate change, it is estimated that a 2°C rise in temperature will drive 20 per cent of all species to extinction and a 4°C rise would eliminate half of all species on the planet. Many species depend on each other and if some are gone many others will follow.

Why is this important? These species and the ecosystems that they constitute provide many ecosystem services that maintain the planet as a good place to live, such as preserving fertile soil, cleaning air and water and providing many resources on which civilization depends. Without them, our future will be seriously compromised. Certainly, as in the past, new species can evolve to replace those that are lost, but only over millions of years and not at a time scale that would be useful for human society.

Food

The next challenge you face on your way out of this valley is the threat of running out of food. We all have to eat and with the rapid growth in the human population and the increasing demand from our rising standard of living, we require ever more food to meet our needs. Some of this has come from intensified agriculture and fisheries, which are now pushing us over several planetary boundaries. We have cleared more and more natural areas for farming, cutting down forests, tilling the soil and applying agricultural chemicals like fertilizers and pesticides. This was behind the green revolution – which saved many people from famine – but at the expense of unsustainable pressures on soil and water supplies. So much land has been converted that not enough is left for other ecosystem services, while overuse has degraded an estimated 38 per cent of all arable land since World War II, as topsoil has eroded away, salt has accumulated from over-irrigation and organic matter has been lost.[29]

We have chemically fixed so much nitrogen for fertilizer and generated nitrogen oxides from fossil fuel combustion that we have overwhelmed the planet's nitrogen cycle, with human sources greater than all natural processes. Our impact on the phosphorus cycle is also reaching its planetary boundary. This not only upsets the natural balances that ecosystems require, but also threatens our future food security. In addition, overfishing the oceans is rampant.

There have been warnings before of mass starvation that technological solutions have pushed back against, but we are running out of options. A shift to a sustainable food system adequate for our human population is still possible,[30] but it will require a radical reorientation in present agricultural systems to restore and maintain soil quality, eliminate the one third of food that is wasted and move to eating plant products lower on the food chain and consuming much less meat. There is still enough food on the planet, but much is lost by feeding it to animals, with the meat and other animal products they produce only worth one tenth of the energy they consume. Eating healthier foods and less or no meat would also have the side effect of ending the obesity epidemic that is sweeping the world at present.

Water

If starvation does not get you, you may die of thirst. While we are not yet close to the planetary boundary for freshwater resources, much of that water is not where it is really needed. Many parts of the world lack enough water for their human populations. Available water is often polluted and mismanaged. Most water use is for irrigated agriculture where rainfall is inadequate. Over-extraction is lowering ground water tables, rivers are running dry and lakes are diminishing and even disappearing when they are not polluted with agricultural chemical runoff. This is threatening the future sustainability of some of the most productive agricultural areas. Climate change will make things much worse, with increasing drought in water-short areas, reduced mountain snowpacks and shrinking glaciers and flooding from increased rainfall intensities. People will have no choice but to migrate out of areas where there is inadequate water to support them.

Energy

The way up the mountain to environmental sustainability gets ever steeper, requiring more energy for your climb. Energy also powers development, allowing us to do more work than we are physically capable of, to live in places that would otherwise be uninhabitable due to heat or cold, to ship goods and travel faster than we can walk, to communicate at a distance and all the other things that make modern civilization possible.

For the last two centuries, we have benefited from the energy subsidy that has come from the exploitation of cheap and energy-dense fossil fuels. However, the easily-accessible reserves are being exhausted and the release of carbon dioxide from fossil fuel combustion is the principal cause of global warming. We must stop the wasteful use of energy through increased efficiency and shift as rapidly as possible to renewable sources of energy that can be sustained on into the future. This will require a complete restructuring of Western civilization – its industry, trade, transport and urban living – within a few decades. This is technically possible, with the challenges largely economic, social and institutional. The short term looks grim, but the long term is bright.

Resources

There are many other resources on which our society depends that are increasingly in short supply. We have mined the most accessible ores of many metals and other minerals. Many forests are logged unsustainably. Costs are rising. The solution is to move to a circular economy in which everything is recycled, with each waste becoming the raw material for another part of the process. Energy is the only thing that degrades beyond reuse, but there is a constant inflow of solar energy that we can tap into instead.

Pollution

If you think the worst might be behind you, you have forgotten about the dangers of being poisoned by pollution. The accumulation of man-made and often toxic chemicals and novel materials threatens us in various ways. While the planetary boundaries have not been calculated for many of these, it is clear that only so much can be released into the environment before they come back to haunt us. One example is endocrine disrupters. Various chemicals used in making plastics have been found to interfere in very small quantities with the hormones in our bodies that are a fundamental part of our internal developmental and communications systems. Many chemicals have been shown to cause cancer, or to interfere with embryonic development, resulting in deformed babies. Others used in agriculture are killing off bees and other useful organisms, reducing soil fertility and polluting water. Micro pollutants and nanoparticles from medicines and cosmetics end up in the environment, causing all sorts of problems. Plastics seemed like a wonderful solution to many problems, until we realized that they hardly degrade and are now polluting our land and oceans.

Another example is radioactive materials. While some of these occur naturally in small quantities in the environment, others like plutonium are entirely man-made and so dangerous that they must be kept isolated from all contact with living things for tens of thousands of years. We have created them in large quantities for nuclear weapons and power generation, leaving an enormous burden for future generations.

One of the major emerging environmental problems is the expense and technical challenge of dismantling all of the nuclear power plants that are reaching the end of their useful life and disposing safely of all the radioactive materials that they have generated along with electricity. This cost has never been included in the price of their electricity and almost no country has found a safe place for long-term storage of the resulting radioactive wastes.

There is also a planetary boundary for air pollution, although for the moment the problems are regional rather than planetary in scale. Europe has made some progress in controlling acid rain from burning high-sulphur fossil fuels. A brown cloud of particulate air pollution from east Asia has been observed extending over India and into the Indian Ocean, reducing the sunlight reaching the surface. Dust from sand storms in the Sahara has fallen across Europe, into the Black Sea and as far as the Caribbean.

Problem of negative messages

A final danger to those hoping to leave this valley is discouragement, and, after reading all of the above, you may well feel discouraged too. Anyone who tries to explain the challenges of environmental protection and sustainability to others, whether to students, the general public or leaders of government, business or public opinion, is faced with a great difficulty. The question that arises is how do you motivate people to positive action when so much of the scientific news is negative?

Do you emphasize scientific objectivity and the lack of certainty about any future trajectory, with all the complexity of explaining risks and probabilities? Do you try for a shock treatment, putting forward the most recent alarming developments and the real possibility of catastrophes on the horizon? In this case, you can easily be discounted as an extremist, again crying wolf, or a Cassandra, forever telling the truth that nobody wants to hear. Do you paint a rosy picture of the wonderful society that could emerge if only people did the right thing and risk accusations of being a utopian dreamer?

The complexities of global environmental change add to the difficulties, with climate change being only one of the interconnected challenges we face. We can no longer think in terms of single problems and single solutions. Human impacts on the environment are now on the same scale as natural processes and globalization has integrated all nations into one world community. As a result, the level of interaction between problems has increased. We must come to appreciate that there is a single global system with many interacting parts and processes operating at multiple nested scales.[31] A small but critical change in one part of the system can have widespread repercussions.

However, the scientific tools are still inadequate to understand and model this complexity and to predict possible consequences. We are even further from having the institutions and management tools necessary to manage and respond to the global changes that have already been triggered. The challenge is made even greater by the widespread scientific disinformation produced by vested interests and the resulting confusion of messages in the media and the public mind. A well-funded anti-science movement, backed by the oil and tobacco industries and fundamentalist religious groups, has filled the media with counter-arguments to climate change and other science-based issues, undermining the credibility of scientists and making it difficult for decision makers to find positions that are both scientifically valid and politically acceptable. It is estimated that almost US$1 billion a year is spent on anti-climate change propaganda and media professionals are employed to skilfully convince politicians and the public that the science is wrong. The difficulties scientists face in trying to communicate the climate change issue range from scientific uncertainty to vicious personal attacks that destroy their credibility. What chance does scientific truth have against more appealing lies and disinformation?

Scientists are also challenged by the moral imperative; that they are expected to remain neutral and objective when the survival of humanity is at stake. Science fails to engage the often more complex and nuanced emotional side of human nature and often thus loses the argument.[32]

The message of science in the short term is basically negative, leading to discouragement and denial. The documentation of the planet's environmental problems and the resulting accelerating global change shows how the inhabitants of the Earth are threatened in fundamental ways. Our economy and way of life are at serious risk, but there is no obvious villain or easy solution. We are both the cause of the problem and the victim.

Forcing scientists into the role of bearers of bad news has contributed to the rise of the anti-science movement. No one likes bad news (except the media, who thrive on it) and the tendency is to shoot the messenger. More seriously in the context of consumer citizenship education, bad news does not motivate positive change, but rather reinforces denial or despair and a feeling of powerlessness before the enormity of the problem. The poor feel like helpless victims and the rich, at best, feel guilty.

It is apparent that scientific information, by itself, is inadequate to motivate action. What has often been missing is the ethical component. Scientific information is necessary but not sufficient to motivate change. It can convince at an intellectual level, but this does not naturally lead to emotional commitment or action. Scientific information needs to be placed in a larger ethical framework of responsibility and solidarity, highlighting the positive social outcomes of uniting in the face of a common challenge.

Acceptance, compassion, cooperation and empathy will produce very different outcomes than aggression, competition, blame and denial. We hold both futures within ourselves and, as we choose whether and how to think about climate change, we are choosing how we will think about ourselves and the new world we are creating.[33]

Climbing out of this valley requires strong values and, above all else, an act of will.

Finding your own path to environmental sustainability

While you are on the difficult and dangerous journey through the valley of environmental crises, you can already do a lot to lay out your own personal path to sustainability. It may seem hard for you to tackle the global problems, but there are many actions within reach and your contribution, along with many others, can make an important difference.

The first foundation of a sustainable lifestyle is contentment with little, getting off the consumer treadmill and focusing on what you really need. Another essential principle is moderation – neither over-indulgence nor complete denial – and appreciating things without excess. The intangible things in life, like social relationships, knowledge, science, culture, art, beauty, contact with nature and the acquisition of spiritual qualities, are much more important than material things once basic needs are met, so why waste time and effort on a materialistic lifestyle and the consumer culture? Ask questions about even the least significant aspects of your life and lifestyle. Experiment with changes that might lighten your footprint upon the earth. Here are some questions to guide you towards sustainable consumption:

- **Usefulness**: Do I really need it?
- **Prevention**: Can I minimize my impact on the environment, health, society, etc?
- **Efficiency**: How was the article produced and delivered, using the least natural resources and energy, with decent working conditions?
- **Quality**: Will it last a long time, avoiding the need to replace it?
- **Solidarity**: Has it been produced and marketed in socially-responsible ways, with equitable sharing of profits?

You are trapped in a material civilization that does not always give you many options. If you want to be of service to society, you may need to make different choices than if you were thinking only of yourself. You will also have different needs and possibilities at different times in your life; for instance, if you have or don't have family responsibilities. You may also be taking many things for granted and have never questioned their relevance for sustainability. The first step you should take is to reexamine your own lifestyle and habits of consumption and to ask if they are consistent with environmental responsibility. Remember that there is no one right way to do things, so you need to make your own choices. The following are some suggestions to help you find your own responsible lifestyle.

Your home, whether a hut in a village, an apartment in a city or a house in a town, can be a significant part of your environmental footprint, but not one that is always easy to change in the short term. When you can, pick your housing location to reduce your need for transport (to work, shopping, school, etc.). Occupy the smallest size residence that meets your needs. It will be more economical to heat or cool, clean and maintain. Aim for an efficient use of space where nothing is wasted. If you are building or buying a residence, choose high quality construction with the best insulation, natural lighting, lowest maintenance and most environmentally-friendly materials possible. Weatherstrip the windows and doors and double the glazing if it is not already done and the temperature extremes require it. If your climate requires heating, consider a low-pollution wood heater or energy-efficient heat pump. Lower the thermostat to avoid excessive heating and do not open the window for a long time when the heat is on; five minutes to refresh the air is sufficient. If you live where air conditioning is common because of the heat, turn up the thermostat and cool as little as possible. Install energy-saving light bulbs and use artificial lighting only when necessary, turning it off when you leave the room. Use energy-efficient appliances and never leave them on standby. Where possible, collect rainwater from the roof for toilet flushing and gardening.

Your energy consumption is both important for wellbeing and a significant source of environmental impact. Your housing should be energy-efficient, as mentioned above and energy for transport is considered below. Choose renewable electricity sources if available locally. Avoid unnecessary electronic devices, electrical appliances and power tools and choose those most highly rated for energy efficiency. While some power tools are hard to replace (such as an electric drill for masonry), many hand tools are just as effective as their powered equivalents for occasional use. For cooking, a microwave oven cooks faster and with less energy than other electric cooking (induction is the most energy efficient form of electric cooking), as is boiling water in an electric kettle. A pressure cooker uses 50–75 per cent less energy than cooking in normal pots. Try cooking several things, or a whole meal, in the same pot. In some places, solar cookers may be practical. Cooking with a biomass fuel (wood or charcoal) is better that fossil fuel (a gas other than biogas; electricity from coal, oil or gas) if it is produced sustainably without threatening forests or vegetation cover. You also need to avoid indoor air pollution and should only use such fuels in rural areas where air pollution is not a problem. Gardening is another area where energy consumption is rising. If you have a lawn, prefer a push mower to a power mower and hand tools or electric garden tools to gasoline/petrol driven tools – which are inefficient, noisy and highly polluting. For clearing brush – if you live in an area where it is possible – try renting a herd of goats or a flock of sheep.

Unless you never leave your immediate surrounding areas, you will certainly require some form of transport. In environmental order of priority: walk if you can; ride a bike or electric bicycle; use public transport – preferably trains, subways, trams or trolley-buses that use renewably-sourced electricity, or buses running on biofuels; join a car-share scheme, or rent a car only when necessary. If you cannot avoid owning a car, choose an electric, hybrid or energy-efficient small model; keep an old car until it wears out (it takes a lot of energy to make a more efficient new car); reserve car use for heavy shopping and going where there is no public transport; carpool for travel to work/

school; and drive as little as possible, combining trips where possible, lowering speed and driving economically. For children going to a school not too far away, join or organize a pedibus to your local school, where children walk together with adult supervision. For your vacations, consider nearby destinations or those reachable by environmentally-friendly transport. Avoid air travel unless necessary for education, work or service to the community. Do not choose driving long distances to avoid all air travel, as the latter may consume less fuel per passenger per kilometre. Carbon credit schemes to financially compensate for air travel or other consumption are a poor second choice to leaving the carbon in the ground; such schemes are better for the conscience than for the environment.

You need to eat, so the environmental impact of your food is important. Calculating the energy cost and environmental impact of food is complicated. For example, vegetables grown locally in a heated greenhouse may require more energy than those shipped from far away and bulk transport may use less energy than your drive home from the supermarket. Favour fresh produce, locally grown and in season, if possible. Prefer simple, wholesome food, with at least five daily portions of vegetables and fresh fruits and avoid snacking between meals. Become a vegetarian, or at least avoid red meat and use only small quantities of meat. Poultry or pork production generates less greenhouse gases than goat, lamb or especially beef. Try to choose fish from sustainable fisheries (MSC) or responsible aquaculture (ASC). If you have a garden, grow as much of your own food as possible. Eat little frozen or pre-prepared food. Avoid bottled water and bottled water-sugar beverages, which are an important source of plastic waste and have high transport cost for little or no benefit. Shop in food stores or supermarkets that feature socially- and environmentally-responsible items and prefer a local farmer's market if there is one near you.

Water is another environmental resource that is now in short supply in many places. Drink local tap water unless it is contaminated. If the water is chlorinated, letting it stand for a day will allow the chlorine time to escape. Run faucets and the shower at low volume with water-saver attachments. Adding an object (a brick or a bottle full of water) to the toilet tank will reduce the volume of each flush. Do not let the water run when it is not immediately needed (like while you brush your teeth). Prefer a shower to a bath; with a low-flow system, it is possible to take a comfortable shower with 10–20 litres of water. Use rainwater to flush the toilet and water the garden, if possible.

You may not think that your clothing has an environmental impact but it does, as it is made, worn and discarded. Pick clothing manufactured in socially and environmentally responsible ways and do not buy more than you really need. Washing clothes uses energy and water. Dry cleaning uses chemicals and produces pollution. Try to minimize the weight and volume of your clothing and wear clothes that require dry cleaning only when necessary. Wash full loads at an economy cycle, with a simple no-phosphate detergent dosed carefully. Hang clothes outdoors to dry and avoid an electric dryer unless there is no alternative. Avoid clothes that require energy-intensive ironing. In choosing a fabric, consider the best balance of criteria: synthetic fibres wash and dry more easily, do not require ironing and hardly wear out (nor do they decompose in landfills); natural fibres like cotton and wool do not come from petrochemicals and are greenhouse gas neutral, but may require more energy and water to clean; much cotton today is produced in unsustainable agriculture, with heavy chemical and energy use and often health impacts on farm workers, while clothing from organic cotton is still hard to find; blends of synthetic and natural fibres may be more practical for shirts, trousers and dresses. Do not discard clothing until it is truly worn out, unless you pass it on to others. Learn to repair things when necessary. Pick conservative timeless styles and keep clothing until it comes back in style.

Household products are another source of environmental impacts. Read the labels and avoid too many chemicals in household products. Aim for simplicity: a simple bath soap without perfume; vinegar-based toilet and bathroom cleaners; zero phosphate detergent for clothes-washing; a simple shampoo. Try to use as little as possible and avoid things that may leave residues or release volatile compounds into the air. Avoid pesticides unless they are essential for health and safety and garden without chemical industry products.

You are living today in a throw-away society, generating unbelievable quantities of waste. Try to produce as little waste as possible, reusing what can be reused, recycling and avoiding throw-away products. Avoid plastics and packaging that are a major source of waste and a global pollution challenge for the oceans. Take advantage of all recycling opportunities available locally and recycle paper, glass, plastic bottles, aluminium, compostable organics, clothing, etc. Return appliances and electronic goods to the stores that sell them, if you can, or take them for recycling, if possible. Dispose of special wastes (batteries, solvents, pharmaceuticals, heavy metals, oil, paints, etc.) properly, not in your general trash. If you have a garden, maintain a compost pile. Use cloth napkins and shopping bags and a minimum of paper towels. Prefer rechargeable batteries for most purposes. Buy recycled products.

The mass media and information technologies owned by powerful business interests often use advertising to try to cultivate an environmentally damaging lifestyle for their profit. Be sensitive to the media and the ways in which they subconsciously manipulate your thinking and emotions. Voluntarily limit your exposure to media messages you did not ask for. Use the Internet and social media in moderation and balance them with real human contact. Your relationship to the natural world is important psychologically and spiritually, so try to keep some contact with nature or living things (plants, animals, an aquarium, walks in a park or the countryside, etc.).

You will find, over time and as you change your desires and habits, that a responsible lifestyle is personally rewarding and also contributes to the betterment of society. It will give you increasing confidence as you face the new challenges of the valley ahead.

3

The Valley of Social Illnesses and the Mountains of Social Justice

The third valley, which ultimately leads to the mountains of social justice, is swarming with people. On its lower slopes are the rural poor, trying to subsist on diminishing resources. The degraded soil, the last few trees for firewood, the long walks to find water and now a changing climate, risk pushing them over the edge into forced migration for survival. The valley bottom is filled with endless slums of people packed together in miserable conditions. On a few hills are the urban castles of the rich, who flaunt their wealth and condemn the poor for not making the effort to succeed as they have done. Crime is rife, riots are frequent and wars occasionally sweep the valley and devastate it before survivors slowly rebuild from the ruins. Crossing this valley is full of risks, requiring humility, honesty and solidarity. Here are some of the challenges you will face in this valley.

Disunity

The basic reason that there are so many people trapped in the bottom of this valley is the lack of unity. Humans are a social species: as infants we cannot survive without a supporting family; children learn to socialize; and every community, culture and nation has its own rules for social cohesion. But as the world has globalized and technology has removed the former barriers between peoples, the old social cohesions of tribe, culture or nation – appropriate to smaller scales of human organization – have lost their meaning. Only by giving first priority to the unity of the whole human race can we appreciate positively the great diversity within it and work for its collective progress.

Without unity, we have a very sick society. Some cling to nationalism and want to keep their nation pure and unadulterated by other peoples. We have seen the consequences of such ideologies in Nazi Germany. In times of insecurity and economic crisis it is easy for those seeking power through populism to fan the flames of xenophobia or a morbid dislike of foreigners – often leading to hate crimes.

These and other forms of prejudice throw up walls between peoples. How can anyone move through and out of this valley when there are so many barriers? Another symptom of social illness is the exclusion or marginalization of certain groups, who are sometimes used as scapegoats and are blamed for everything wrong in society to distract attention from the real causes of the problems.

Inequality

Another obstacle in this valley is extreme inequality. While the world economy has created unimaginable wealth, this has become more and more concentrated in the hands of a few, increasing levels of social inequality. Sometimes these differences are institutionalized and perpetuated as social classes or castes that children are born into and cannot leave. Again, these are social barriers that prevent people from fulfilling their potential and truly contributing to society.

The extreme differences in wealth between countries and the

failure to resolve the challenges of development among the poor, lead to wave upon wave of migration as people try to escape poverty and look to wealthier regions for a better life. To this is added the flood of refugees escaping war and persecution and soon climate change will displace many millions more. Today this is seen as a social crisis; but, people have always migrated and national frontiers are a relatively new invention. Immigration, in fact, is good for an economy.[1] The American economy, for example, was built by immigrants. Rather than waiting for humanitarian crises to force migration upon us, society should anticipate the need for people to move as part of social justice and manage the process to everyone's benefit.

Social breakdown

Another risk in this valley of social illnesses is the breakdown of family structures and relationships, in which children are often the principal victims. There is a similar breakdown in any sense of community, with so many people feeling lonely – even within a busy city. Some of the symptoms are rising levels of crime and insecurity. Tellingly, this is most obvious with rising levels of wealth, when wealth ought to be increasing wellbeing rather than reducing it.

The largest scale of social illness is expressed as war and terrorism, when one group or nation sets out to destroy as much of another as possible. What a contradiction, when everyone dreams of peace and security! Today the rot has spread so far that we see widespread failures of governance and even failed states.

None of this is inevitable or beyond our control. All human institutions were designed and built by people, so we can change them if we really want to. But this requires both more of a systems approach and the transformation of people, with challenges you will face in a later valley. For the moment, let us address social cohesion and then whether we are naturally aggressive and violent or capable of rising above such impulses and behaviours.

Social cohesion

One key to escaping from this valley is social cohesion. Recent research, applying the techniques used to model the rise and fall of animal populations and to explore similar processes in empires and civilizations, has analysed what makes empires grow and then collapse – looking at Russia, ancient Rome, Islam and Medieval Europe, among others.

Societies on a frontier between two very different cultures are subject to stresses that force them to build social cohesion to resist enemies. The strength that comes from cohesion, effective organization and a willingness to sacrifice for the common good allows such societies to expand into empires. The success of an empire, however, contains the seeds of its own downfall. The wealth of a growing population and improving technology produce a successful elite, until excess population growth allows increased exploitation of labour and an overshoot of food production capacity – in which the poor suffer and the elite continue to live well. A generation later, the excessive concentration of wealth leads to conflict among a too numerous elite over a shrinking resource base. The young rich start falling into poverty and revolt, and the civilization loses cohesion and collapses, perhaps through several cycles.[2]

This view is a largely negative one, showing that it takes wars to build empires and civil wars to destroy them. Within such a framework, wars are what build a spirit of cooperation in society and are reflective of cultures with the highest levels of cooperation.[3]

It is worth asking whether forces other than war and the constant threat of an enemy on a geographical frontier can bring about social cohesion. The rise of a new religion, for example, can create another kind of cultural frontier, with those sharing the new values building social cohesion as they work to transform society. A highly cohesive social and spiritual movement could overcome the negative forces around it and expand rapidly into a global civilization, as did Islam.

The mountain range of justice that rises out of this valley is founded on social cohesion. This comes not from constant external threats but from an inner spiritual force for unity in diversity. Such social cohesion can build a civilization able to achieve sustainability in its use of resources, prevent the excessive concentration of wealth in an elite and thus rise above the cycle of decline and fall that has characterized past civilizations – or at least slow the cycle to the millennial span of religious revelations.

Altruism and cooperation

Another barrier to leaving this valley is the assumption, most obvious in today's dominant neoliberal economic paradigm, that people are inevitably selfish, aggressive and competitive, only looking out for their own self-interest. Many people have adopted these values because they believe these are natural human impulses. Yet much recent research shows that this is not the case.

Group selection favours altruism, as it enhances the strength and competitiveness of groups and it has advanced during human evolution by natural selection at the group level. While selfish individuals defeat altruistic individuals, groups of altruists defeat groups of selfish individuals. We are basically tribal and seek out groups and these groups favour trust and virtue. We automatically feel empathy for the pain of others and are prone to be moral, to do the right thing, to hold back, to give aid to others – sometimes even at personal risk – because natural selection has favoured those interactions of group members benefitting the group as a whole.[4]

The emotions with which we condemn others, like contempt, anger and disgust, prompt us to punish cheaters. The emotions with which we praise others, including gratitude and elevation, moral awe or being moved, encourage us to reward altruists. The emotions that allow us to share the suffering of others, such as sympathy, compassion and empathy, prompt us to help a needy beneficiary. The self-conscious emotions of guilt, shame and embarrassment incite us to avoid cheating or to repair its effects.[5]

Other research also demonstrates that cooperation wins out over cheating. In the classic prisoner's dilemma, the cheater who sells out his partner tends to win, but as soon as a social dimension is added where people refuse to cooperate with cheaters, a cooperative approach is more effective. Altruism is at the core of social justice and cohesion.[6]

More broadly, among those things that contribute to happiness is the happiness that comes from leading an ethical and virtuous life.[7] Of course, this is nothing new, as religions have been saying as much for thousands of years. It is our materialistic society that has forgotten our fundamental purpose to cultivate ethical and spiritual virtues like altruism and to rise above the selfishness of our animal nature that normally we should grow out of in childhood.

Putting social justice into action

Now that you can rise above the misconceptions about social relationships and accept that change is possible, you can start your climb out of this valley up to the mountains of justice on the path of social action. This is not to say that we know all the answers and the path will be easy. It is a difficult climb; but the guide below is a synthesis of the processes that need to be followed.[8]

If our goal is social justice, we need clear concepts of society and social progress. There is general agreement today that social progress requires a fundamental transformation in society,[9] but the nature of that transformation is less clear. First, we need to agree on fundamental issues of existence, such as the nature of the human being having both material and spiritual realities and the purpose of life. We can then acknowledge that civilization has both a material and a spiritual dimension and that humanity is on the threshold of its collective maturity, building a world society. As in any time of transition in a complex system, there are destructive and constructive forces operating in the world which serve to propel humanity towards its full maturity. All social relationships need to be recast in this context, transforming simultaneously both human consciousness and the structure of social institutions. This means analysing and rethinking concepts such as power, authority, individualism, personal comfort, selfless service, work and excellence.

Where injustice has often come from abuse of power and arbitrary authority, new institutional models of govenance are needed, free from individuals seeking power. Individual initiative needs to be balanced by social responsibility. Service to others and to all of society should become the highest ideal and motivation for work, rather than accumulating wealth and seeking personal comfort. The focus of excellence should shift to intellectual, scientific, artistic and spiritual attainment. Individuals motivated by altruism and selfless service will become the best instruments for attaining social justice.

Social action

Transforming the world so that everyone can escape from this valley cannot just be an intellectual exercise. It must lead to social action and this is where each of us can find a place. Social action is not some form of development aid which can destroy dignity or create dependency, nor is it experts helping the ignorant. It should be participatory, empowering us to take charge of our own development according to our own priorities and accompanying one another as we build community capacity and learn together. Social action needs strong spiritual foundations, with material advancement balanced by spiritual motivation and a vision of what the community wants to achieve together. The community develops the skills of consultation, action and reflection as it grows through its own learning. For those of you who are interested in committing your life to this, the box provides a detailed guide to effective social action.

Guide to social action

To start with, it is necessary to understand the goal of social action in a new way. It is not just some kind of social or economic development. It is not just the transfer to all societies of the ideological convictions, social structures, economic practices, models of governance and the very patterns of life prevalent in certain highly industrialized regions of the world, or the mere consumption of goods and services and the naive use of technological packages. It must avoid dividing the world into 'developed' and 'underdeveloped', where development aid creates a relationship of charity and dependency. Social change should not be a project that one group carries out for the benefit of another, or where an expert imposes their professional perspective.

Social action must be participatory, involving the people themselves in the generation and application of knowledge, eliminating dualities such as 'outsider-insider' and 'knowledgeable-ignorant'. The aims and methods used should empower rather than destroy the dignity of those involved. The approach should give due respect to local autonomy, while accepting to be part of a larger whole; attach importance to material means, without allowing them to become instruments of control; provide for the flow of knowledge, without introducing paternalistic attitudes; and strengthen capacity in individuals, without any regard for their economic status.

If you are to avoid contradictions, you need to become increasingly aware of the thinking and pressures of your social environment. While you can freely draw insights from the philosophies, academic theories, community programmes and social movements around you and keep current with the technological trends that influence progress, you should remain watchful that your values and spiritual principles are not bent into conformity with this or that ideology, intellectual fad or fashionable practice. You need to measure prevalent approaches, ideas, attitudes and methods with reference to their underlying values. With this capacity you can uncover the aggrandizement of self, so often lying behind initiatives that are nominally concerned with empowerment, discern the tendency of certain development efforts to foist upon the poor an entirely materialistic worldview, perceive the subtle ways in which competitiveness and greed can be promoted in the name of justice and prosperity and ultimately abandon the notion that one or another theory or movement — which may fleetingly acquire some prominence in the wider society — can provide a shortcut to meaningful change.

Everything has to start with community building, since social action comes naturally from stirrings at the grassroots rather than something imposed from outside. The main aim should be building the community's own capacity, starting at a modest scale and only growing in complexity as available human resources allow.

Learning is central to action, since there is no one 'right' way and every situation is different. The aim should be a spirit of humble learning, characterized by constant action, reflection, consultation and study, re-examining visions and strategies. As tasks are accomplished, obstacles removed, resources multiplied and lessons learned, modifications are made in goals and methods. Learning is like the growth and differentiation of a living organism, continually advancing towards the fulfilment of its purpose while adapting to its environment. By combining this process of learning at the local level with the access to knowledge that is now available through electronic media and larger social networks, the approach avoids categorization as either bottom-up or top-down and becomes one of reciprocity and interconnectedness.[10]

Bringing a spiritual dimension to social action

The purpose of civilization should be to foster true prosperity, with its spiritual and material dimensions, among the diverse inhabitants of the planet. It is therefore vital to achieve a dynamic coherence between the practical and spiritual requirements of life. Effective community action needs strong spiritual foundations.

You will need to keep in mind both the material and spiritual dimensions of the life of a community and give due attention to both scientific and spiritual knowledge. These two sources of knowledge tap into the roots of motivation in individuals and communities, so essential in breaking free from passivity and enable them to uncover the traps of consumerism. For example, scientific knowledge helps the members of a community to analyse the physical and social implications of a given technological proposal – say, its environmental impact – and spiritual insight gives rise to moral imperatives that uphold social harmony and that ensure technology serves the common good.[11]

Participation

Participation is another important characteristic of social action. You will want to ensure that you involve a growing number of people in a collective process of learning, focused on the nature and dynamics of a path to material and spiritual progress in the community. Everyone can engage in the generation, application and diffusion of knowledge, which is a potent and indispensable force in the advancement of civilization. In addition, the application and propagation of existing knowledge is invariably accompanied by the generation of new knowledge, including the insights acquired through experience.

In many ways, the immediate goal of a social action to produce a beneficial result is less important than the accompanying rise in the capacity of the participants. While they acquire knowledge and skills in a specific area, they strengthen their capacities to foster unity in diversity, promote justice, participate effectively in consultation and accompany others in their efforts to serve humanity. Even the smallest group of local individuals must be able to maintain a consultative environment based on honesty, fairness, patience, tolerance and courtesy.

A process of community development needs to reach beyond the level of activity and concern itself with the culture of the community. Social action can become an occasion to raise collective consciousness of such vital principles as oneness, justice and the equality of women and men; to promote truthfulness, equity, trustworthiness and generosity; to enhance the ability of a community to resist the influence of destructive social forces; to demonstrate the value of cooperation as an organizing principle for activity; to fortify the collective will to act; and to combine practice with spiritual principles.[12]

Methods of social action

The following concepts can suggest some of the methods that can be adopted in support of social action. If a group is consulting together in the collective investigation of reality, it helps to create an atmosphere which encourages detachment from personal views and listening to others with an open mind in a collective search for the truth. Give due importance to valid empirical information, without mere opinion being raised to the status of fact. Conclusions should correspond to the complexity of the issues at hand and not be broken up into a series of simplistic points. The results and conclusions of the consultation should be presented in precise and dispassionate language. The method is important, as progress in every area of endeavour depends on the creation of an environment where everyone contributes and supports unified action.

One basic skill is reading society and formulating a vision. Every social action responds to some understanding of the nature and state of society, its challenges, the institutions operating in it, the forces influencing it and the capacities of its peoples. Reading society does not require formal studies or a review of every detail of the social reality. Conditions need to be understood progressively, both from the perspective of the purpose of a social action and in the context of spiritual principle and a vision of humanity's collective purpose. This reading should be continually refined using the methods of science.

According to your reading of society, you will need to form and refine a vision of your work within your social space. This vision is not simply a set of goals or a description of an idealized future condition. It has to express a general idea of how goals are to be achieved, the nature of the strategies to be devised, the approaches to be taken, the attitudes to be assumed and even an outline of some of the methods to be employed. Such a vision of work is never complete; it has to become more and more precise, be able to accommodate constantly evolving and ever more complex action and attain increasingly high levels of accuracy in its operation.

Consultation is another essential tool for learning in action. Whether analysing a specific problem, attaining higher degrees of understanding on a given issue, or exploring possible courses of action, consultation may be seen as collective search for truth. Participants in a consultative process see reality from different points of view and, as these views are examined and understood, clarity is achieved. In this conception of the collective investigation of reality, truth is not a compromise between opposing interest groups. Nor does the desire to exercise power over one another animate participants in the consultative process. What they seek, rather, is the power of unified thought and action. A consultative spirit should pervade the interactions of those engaged in social action, of whatever size and complexity and the population they serve, so that the aspirations of the people, their observations and ideas are ever present and are consciously incorporated into plans and programmes.

Action and reflection on action form another critical methodology. At the heart of every endeavour is consistent, systematic action. Action, however, needs to be accompanied by constant reflection to ensure that it continues to serve the aims of the endeavour. Learning depends on structured reflection woven into a pattern of action, through which questions can emerge and methods and approaches can be adjusted. It is important to emphasize the spirit with which action is undertaken, which should include purity of motive, rectitude of conduct, humility, selflessness and respect for human dignity.

In general, a challenge for any social action is to ensure consistency among the explicit and implicit convictions which underpin an initiative, the values promoted by it, the attitudes adopted by its participants, the methods they employ and the ends they seek. This requires consistency between belief and practice. A deep-seated recognition of the oneness of humanity should prevent all efforts from fostering disunity, isolation, separateness or competition. An unshakeable conviction in the nobility of human beings and their capacity to subdue their lower passions and develop heavenly qualities should serve to protect against prejudice and paternalism, both of which violate the dignity of people. An immutable belief in justice should ensure that resources are allocated according to the real needs and aspirations of the community, rather than the whims and wishes of a privileged few. The principle of the equality of women and men should open the way not only for women to assume their role as protagonists of development and benefit from its fruits, but also for women's experience to be given more emphasis.[13]

Community action

Building on the tools as described in the box, a new field of service is open to you as you begin to climb the mountains of social justice. The best place to start in creating social justice is your own community. For example, if you are young, you might begin by learning to accompany groups of pre-youth (11–14 years old) to strengthen their characters and help them discover the satisfaction that comes from acts of service to the community. Or perhaps you could organize a children's class in your neighbourhood to teach ethics and virtues to the next generation. The Bahá'ís have excellent training programmes for this that have proven their worth around the world. They also have materials for study circles, appropriate for the whole community, that can build a common understanding and purpose and motivate people to become actors in their own community development. Devotional gatherings of those of all faiths and those of none can provide a spiritual foundation for community action.

A community that has a shared spirit of solidarity, that reflects on its progress and consults on the next steps forward, that respects everyone and leaves no one behind, is the best insurance against any of the many problems that may threaten society in the years ahead. Even if the larger problems of the world seem beyond reach, community action can go far to reach the peaks of social justice and open the way to the further challenges in the valleys ahead. It is in this direct way, person to person at the neighbourhood or community level, that you can start building the unity that the world so badly needs. You do not need any special talents, training or experience, but can learn as you go along.

This is also the solution, at the most basic level, to the challenges that increasing migration is bringing to the world. When newcomers arrive and people need to learn to live together in harmony, what better way than by studying together about the principles of unity in diversity, sharing devotions across different faith traditions and consulting on the common needs of the community while building mutual understanding. If all the children learn and play together in neighbourhood classes and the junior youth organize projects of service to the community, regardless of their origins, then the foundation is laid for lasting unity in the community. This is the best expression of social justice, wherever you live.

So, equipped with your new understanding of the power of social justice and with the tools of social action, you are now ready to go beyond the mountains of justice and enter the next valley on your journey of self-realization.

4

The Valley of Economic Crisis and the Mountains of Transformation

The valley of the economy is a desolate place full of smoke and fire, pitfalls and dangers. The well-beaten paths tend to lead to wars and destruction. There are deep chasms into which the economy can unexpectedly fall. It is easy to be drawn to superficial distractions and to stop moving, becoming a passive consumer of worthless things and staying forever in the dark bottom of the valley, far from the light. There are temptations on every side, sold persuasively by well-planned marketing. To escape this valley, you need strong willpower and a vision of a higher and more rewarding reality – one that you will find up in the distant mountains.

The financial system and the growth paradigm

One source of the smoke and fire in this valley is the economic free-for-all of a globalized economy, where there are no mechanisms for global economic management or common regulations for multinational corporations. Beyond national borders, anything goes. Since the end of the cold war and the collapse or transformation of most communist systems, the free-market economy has spread around the world and created wealth on a scale unimaginable in the past. It demonstrates the triumph of deregulated enterprise and innovation, including financial innovation and confirms the dominance of the United States and its consumer culture as the world's largest economy and the model to follow. Economists convinced by this model have been the drivers behind most government policies for the last few decades.

The western economy and its materialistic values have been exported to every corner of the world. The material success of this system has made it impervious to critiques that it might be socially or environmentally unsustainable in the long term. Growth was the unchallengeable goal of business and government and suggestions that there might be limits to growth were derided and ignored. Experts in the scientific and environmental communities have long suggested that humans might also be vulnerable to overshoot and collapse. They have only recently begun to attract attention after decades in the wilderness of indifference (as you have seen in the valleys you have already crossed). Evidence for climate change, like melting ice caps and the rapid rise and general instability in food and energy prices, signalled even to economists that something was wrong in the system.

Fire burst out in the economy with the collapse of the banking system in 2008, after the American sub-prime mortgage scandal and its propagation throughout the world economy. This provided a sudden challenge to the economic paradigm.[1] Sophisticated mathematical models for managing financial risk proved incapable of managing greed. The triumphant free-market economy went up in flames. Stock markets plunged and unemployment surged. Price volatility for fossil fuels and food destabilized national economies and plunged millions more into poverty and hunger.

As the world recession deepened, government intervention on an unimaginable scale was required to head off a complete collapse of the financial system and economy. Money was poured into the economy, like water into a building on fire. The attempt to restore confidence in the banks and to restart lending saw debt increasingly being transferred to governments – on the assumption that no one would worry about governments' ability to repay their debts. Social unrest increased as workers asked why the rich got all the handouts while they themselves lost their jobs. Understanding the complexity of the situation, the head of the European Central Bank, Jean-Claude Trichet, said, 'We live in non-linear times: the classic economic models and theories cannot be applied and future development cannot be foreseen'.[3]

The consequences of the fires still smouldering in the economic system are far from evident. Augusto Lopez-Claros, a distinguished economist, warned in December 2008 that the real danger was not a deepening recession, but the possibility that massive government intervention might produce a partial recovery. Leaders would be convinced that they had avoided the worst and would not recognize that the economic system was broken and in need of fundamental reform. Having used every weapon in their armoury, there would be nothing left with which governments could respond to in the next crisis.[4] This is exactly what has happened.

The next fire to break out could be among national governments, with a number of countries already on the verge of bankruptcy.[5] And the fire seems to be spreading. A loss of confidence in governments could bring down the whole global monetary system on which the world economy is based. Perceptive speculators are fanning the flames, pocketing enormous profits, while betting against the very financial system that created them.

In the lead-up to the Paris Climate Change Conference in December 2015, the Governor of the Bank of England and chairman of the G20 countries' Financial Stability Board, Mark Carney, warned that climate change might make the world's stock markets and banks unstable and lead to a financial crash. A fear of stranded assets, leading to panic selling, could cause a plunge in value of shares in fossil fuel companies and industries that produce a lot of carbon dioxide, amounting to one-third of stock market assets. Banks might become unstable because the billions of dollars in loans they have made to fossil fuel companies might not be repaid. He further explained: 'Our societies face a series of profound environmental and social challenges. The combination of the weight of scientific evidence and the dynamics of the financial system suggest that, in the fullness of time, climate change will threaten financial resilience and longer-term prosperity. While there is still time to act, the window of opportunity is finite and shrinking.'[6] Despite his efforts since to reduce this risk, it is a conflagration that could break out at any time.

A failure of values

To take the metaphor a step further, the fire in the economy resembles some of the classical descriptions of hell, where greed has replaced need as the driver of growth at all costs. The current economic paradigm depicts growth as measured with GDP as the only way to improve society. Economic growth is seen as the solution to all our problems.

There is a logic to the economists' attachment to growth. A system based on borrowing and debt cannot survive otherwise. A growing economy will create new wealth to reimburse current borrowing and interest. Without growth (or the shadow of growth that results from inflation), there is no way that debts and accumulating interest can be repaid. The whole system collapses like a house of cards.[7]

To find a way out of this valley, you will need to question the basic values on which modern society has been based, where success is measured by growth in the economy, in business and in politics and any leader who does not bring growth in GDP, profits or power is quickly replaced. The growth paradigm may have been a necessary response to an expanding population, an increased energy supply and growing resource exploitation, on which technological advances have been built. However, the United Nations estimates that the world population will stabilize in the middle of the 21st century and it already has done so in many wealthy countries. The decline in fossil fuel reserves means the end of the cheap energy subsidy on which industrialization, trade and intensive agriculture have been based. In a heavily-exploited world, it is difficult to see where further significant growth in natural resources can come from. Economic growth as we have known it cannot continue, except to respond to the needs of the poor.

Avoiding overshoot and collapse means abandoning the growth paradigm for a model based on balance, efficiency, equitable distribution, optimal sizes, renewable energy and closed materials systems – in what is sometimes called the circular economy. As in nature, the wastes of one part of the system become the inputs to another, so everything is recycled and nothing is simply thrown away. This implies a revolutionary change in economics, politics and society. This change will not come easily, but it is the only way to put out the fire.

The continuing difficulties in the financial system have changed the diplomatic landscape, put new issues on the table and broken major resistances to public intervention and international action, but not enough to create a consensus on ways forward. They have demonstrated how fundamental trust and confidence are to any human system or institution, including banking, business, government and diplomacy; how easily they are lost and how difficult they are to restore. They also show how all the different problems are interrelated and that solutions to all of them have to be considered together, challenging the traditional approach of dealing with different issues in separate arenas and seldom making the linkages between issue-specific agreements and regimes.

To find a way forward, both to address the immediate causes and consequences of the financial problems and to explore the possibilities to move in the direction of a more sustainable economy, we need to consider some key questions:

What are the wider consequences for people and for a planet already facing ecological crisis, of the near-failure of the banking system and the enormous financial effort by governments to prevent the total collapse of the economic system?

Austerity has become the key issue for governments as they try to rebalance their accounts. Budget cuts generally include environmental and social programmes that should be cushioning the shock and preparing for the future. Short-term urgency squeezes out long-term necessity. There is a lack of capital to invest in other priorities, unless they can be shown to meet also the requirements of the economy.

Will the rise in social instability make the transition to a more sustainable system more difficult, or will it facilitate it?

There is a growing sense of injustice among workers and the poor, that so much help has gone to the rich while rising unemployment and prices impact the poor the most. The middle classes stagnate and their children look to a future worse than their parents. Pressures on governments to increase protectionism are growing, threatening to put a brake on global exchanges.

This defensive turning inward could make it more difficult to find global solutions to climate change and other environmental problems.

What are the underlying causes of the financial crisis and subsequent stagnation?

Are there failures of conception in the way the economic system works? Are there systems errors, in which each part makes sense but they interact in ways that are not anticipated? Are the failures not in the mechanisms but, more fundamentally, at the level of ethics and values? It could well be that the whole Western capitalist model is as fundamentally flawed as the communist model, with a focus on shareholder value rather than value to society – investing for the highest return, borrowing rather than saving, encouraging excessive risk taking and speculation and seeing ever-rising levels of debt as 'normal'. The amounts that governments are borrowing today to finance their interventions in the economic system through quantitative easing are so unimaginably large they question the meaning of debt and money for ordinary people.

Is finance the central problem or only the symptom of something deeper?

The same process of debt accumulation and living beyond our means can be seen not only in the financial system but in social and environmental systems, all of which are showing increasing instability and signs of reaching their limits. There is a danger of converging financial, environmental and social bankruptcies, which some experts say could lead to the collapse of civilization. The lessons from the financial crisis may be equally valid for other dimensions of society and the environment. This is a further argument for abandoning the growth paradigm and replacing it with another that is more aligned with sustainability. At the moment we seem to be trapped in an economy heading for disaster and unwilling or unable to change course. Blind optimism in technological solutions too closely resembles the optimism in the financial system not so long ago.

How do we deal with an economic system that is so dependent on consumption to create employment, when environmental sustainability seems to require limits on or even a reduction in consumption? How do we redistribute wealth so that poverty can be reduced, implying even greater changes in lifestyle among those who are better off?

Even a slight reduction in purchasing power brings people into the streets and becomes politically destabilising. We are trapped between an objective reality and a political impossibility. It is no wonder that this valley is full of smoke and fire.

Limits to future growth

The catastrophic fires sweeping this valley should not be a surprise. Economists have had ample warnings. You saw earlier how the report for the Club of Rome on *The Limits to Growth* presented a computer-generated scenario showing that 'business as usual' would lead to the collapse of civilization in this century. One of the authors of the original report and its update forecast in 2012 the most probable future for the next 40 years by projecting the trends observed since 1972.[8] There is a natural tendency, reinforced in democratic systems and in the capitalist economy, to always choose the least-cost, short-term solution. We only change when we have to and no more than absolutely necessary, so the result is always too little, too late. The population will stop growing, but only because fertility rates will drop in cities.

We shall pursue GDP growth because it is the only way to create jobs and distribute wealth, but growth will slow down, only doubling by 2052 and most of that growth will be in China and the emerging economies. The rich countries are reaching the limits of productivity increases, so their growth will stop and, in the United States, probably decline. More economic effort will have to go into correcting environmental damage and rebuilding after natural disasters are triggered by climate change, so we shall have to work harder just to stand still. The beauties of nature and undisturbed ecosystems will disappear. There will be enough resources to meet the demand, but not the need; five billion people will continue to live in poverty and a billion will starve if nothing is done to address the extremes of wealth and poverty. Inequity in the rich world will increase, producing more social instability. The young will rebel against their elders – who expect to live comfortable retirements while leaving their grandchildren to pay the price for their excesses. The market will not solve these problems and democracy will fail to align economic and social interests. There is a brief mention of wildcards that could upset this forecast, including a financial meltdown, a revolution in the USA and a generational rebellion.[9] These predictions are a good estimate of where the world will go, at the bottom of this valley, if there are no major changes. The perspective is not encouraging. The fires will continue.

A much better future is technically possible, requiring a shift of only two per cent of labour and capital. But, this is slightly more expensive, so we do nothing and face disaster just over the horizon. This rather pessimistic analysis is built around five central issues: capitalism leads inevitably to extremes of wealth and poverty; economic growth produces over-consumption; democracy is too slow for the changes that are necessary;

intergenerational harmony will fail; and the climate will become increasingly unstable.[10]

This scenario is based on our present materialistic value system and projects it into the future without any significant change. It does not consider that there may be other forces at work in the world and that the inevitable decline in our present disfunctional system can open the way for the birth of a new global civilization – one that is founded on new values.

Other thinkers quoted in this forecast to 2052 suggest solutions including: changes in corporate values towards more responsibility; the potential for open and collective innovation with new technologies; the possible evolution in human values reaching peak with a generation that is more educated, connected and spiritual; waking up with the spiritual strength and rational clarity for repentance and conversion; and taking the next step in human cultural evolution to a higher level of organization with consultative decision making and the acceptance of our spiritual reality.[11] It is clear that many other people see constructive ways out of this valley.

These are the economic challenges for all of us today. You are our best hope if reading this empowers you to use your potential for sacrificial service, innovation and collaboration to transform the system from the bottom up. Forty years is enough time for two generations to bring about exciting change.

Scoping solutions

You should not feel discouraged by the gloomy economic scene described above. Many people are looking for solutions and making progress. There are encouraging efforts at all levels in society. If the present financial system does collapse, there are alternatives waiting to be built on the ashes of the old system.

One dimension of the way ahead that is being pushed

strongly is the rapidly developing concept of the 'green economy', defined as an economic system that preserves and restores ecosystems as the backbones of economic and social wellbeing and is essential for poverty reduction while simultaneously creating new industries and employment. Environmental industries using clean and efficient technologies and sustainable agriculture would serve as major engines of wealth and job creation and poverty reduction. The aim is to find combined 'win-win' solutions for the economy, human wellbeing and the environment. This has been a major theme for UN Environment (UNEP) and at the UN Rio+20 Summit in 2012.

UNEP[12] has identified the priority green economic sectors as:

- Clean and efficient technologies, including renewable energy technologies and rural energy access.
- Biodiversity-based businesses, including agriculture, forestry, marine, nature-based tourism, etc.
- Ecological infrastructure, including nature reserves, protected areas, watersheds, etc.
- Chemicals and waste management, including waste reduction, recycling and reuse.
- Low carbon cities, buildings and transport.

The economy also needs to respond to the challenges of climate change. Imagination and innovation will be required, as 'business as usual' is no longer a realistic option. The coordinated financial stimulus required to rebuild the world economy can either lock us into a vulnerable fossil fuel-based system or accelerate investment in a low-carbon transition – creating jobs, technological innovation and market stimulus and using public funds to trigger much more private finance.

We need an unprecedented multi-stakeholder collaboration to link the climate and economic agendas, responding to 'the shared desire to deliver climate security, energy, food and water security, economic security, equity between rich and poor through enhanced capital and technology flows, all through the creation of a package that promotes economic growth by decarbonizing the world economy'.[13]

The Global Commission on the Economy and Climate[14] has shown that a new climate economy based on renewable energy sources would create better growth with more employment and a higher GDP than 'business as usual' with a fossil fuel-based economy, while also protecting us from climate change. There is no economic reason not to make a rapid transition now.

However, most present efforts to address the economic crisis have emphasized relaunching consumption and maintaining sectors, such as the automobile industry, in the hope of returning to the consumer economy, with only marginal efforts to invest in more 'green' alternatives. Too many vested interests want to protect the status quo. There is a danger that the 'green economy' would just be consumption with a new name, when the real need is to place limits on consumerism. The debate on how to do this is only on the fringes and the resistance to change is very strong. Furthermore, all of these proposals are still within the economists' growth paradigm, which is unsustainable – as growth cannot continue forever in a finite system. The consumer society is overdeveloped and the costs of unsustainability are rising. Some people talk about 'degrowth' in wealthy countries. What is needed is another economy, emphasizing the social rather than the material forms of development.

Governance

Putting out a large fire requires a coordinated response by many actors and this means effective governance at the same scale as the problem. The economy has globalized, but economic management mechanisms are largely national. The system of governance based on national sovereignty has great difficulty in addressing global problems effectively, as illustrated by the slow pace of intergovernmental action on climate change.

At the same time, the diversity of situations around the world requires a nested set of levels of governance to keep decision making close to the level of action and to encourage innovation and local adaptation. We need to evolve institutions of governance at all of the scales of the problems we face in global change.

There is presently a strong prejudice in many quarters against global government, often reflecting the ideological or political position that government is essentially inefficient and bureaucratic and the less we have of it the better. Yet, effective government is an essential component of any civilized society. Even the business community recognizes this. Europe has pioneered the creation of institutions at a regional scale in complement to national governments and this example, however imperfect, will be equally appropriate as we consider how to deal with global change.

As long as there are extreme differences in wealth and level of development between and within countries, it is necessary to apply the approach of common but differentiated responsibilities, with each country adopting its own implementation strategy within the larger framework. However, all must respond to the required fundamental transformation in the dominant economic paradigm and its consumer lifestyles. How to organize the transition and cushion its negative impacts are major challenges. In addressing climate change, for example, the adoption of deep emission cuts in the north, on the principle of equal per capita emission rights, will produce significant transfers of investments from north to south.

The financial collapse and its aftermath have underlined the need for a systems approach integrating all the issues. Science and technology have united the world in information flow, finance, trade, migration and environmental impact. Each part affects the others, so they must be understood together in all their complexity. Yet our society is not well structured to deal with complex multidisciplinary global problems. Governments are divided into ministries or departments and the academic community into disciplines. There are strong pressures for increasing specialization. Perhaps a first step would be to encourage a specialization of a generalist, integrator, or systems manager, with an accepted role in bridging disciplines. Education also can include systems thinking and integrative approaches as part of general education. The attentive consumer needs this capacity to assess and integrate many kinds of information, how much more so decision makers in business and government.

In education and public awareness, these issues should be treated proactively. Young people, at least, should feel comfortable with integrated thinking and consider it normal that there be institutions of governance at all levels. There should be among the foundation principles an acknowledgement of our world citizenship. However, more is needed to motivate action. The challenge is to be both scientifically objective and realistic about the threats and risks of global change and its implications for the economy, society and consumer behaviour, while also inspiring hope and a desire to act and seeing the necessary sacrifices in a positive light.

The engagement of business is also critical. The first encounters between business and environmental interests in the 1960s and 1970s were negative. Industrial pollution had caused great damage and the costs of pollution control and clean-up were considerable. Governments instituted increasing levels of environmental regulation in the public interest against the strong opposition of the business community. This resulted in the general assumption that environmental regulations reduce profits and damage competitiveness. Today, that assumption persists, particularly in government. Many governments continue to believe that in order to stimulate the economy and create wealth, government regulation and interference with business should be minimized or eliminated.

But the reality is different. One of the surprising results of a survey of thousands of business leaders in 104 countries,[15] was the demonstration that business leaders considered that good governance – expressed in the enforcement of strict environmental regulation – actually increased their competitiveness.[16] The countries where business thinking on environmental and social responsibility is most advanced are also those with some of the most advanced and competitive industrial economies. Having strong regulations allows companies to compete in meeting their regulatory obligations, giving a competitive advantage to companies that innovate and increase their efficiency in environmental performance, as well as opening up new market niches for environmental services.

The ethical component

Each culture and nation has evolved and institutionalized its own ethical framework within the context of its religious, cultural and philosophical heritage, representing the ethical consensus of its society. However, rapid globalization has taken the ethical issues and challenges to the planetary level, for which the self-contained national sets of values are poorly adapted.

In response, there have been efforts at the intergovernmental level to adopt declarations of ethical principles, such as the 1948 *Universal Declaration of Human Rights*[17] and subsequent declarations and conventions, the *Stockholm Declaration*[18] in the environmental area, the *Rio Declaration*,[19] and other similar documents. Civil society has tried to go further with texts like the *Earth Charter*.[20] However, it is not always easy to translate these general principles into guides for practical action, whether at the governmental level or in the behaviour of each individual. Where self-interest and ethical principles are in conflict (as they frequently are), self-interest (including national self-interest) has too often won out.

An ethical or moral framework of what is right or wrong underlies most systems of human organization, whether in traditional cultures, religions or legal systems. Central to all these frameworks is the necessary balance between individual freedom or satisfaction and the collective wellbeing of society. A mature citizen with confidence in the government will voluntarily accept to obey laws and pay taxes in the common interest.

One of the prominent features of the last century was the rise of materialism as the dominant value system, expressed in the prominence of economic thinking in government and business (Adam Smith's 'invisible hand of self-interest'[21]) and the rise of the consumer society focusing on individual satisfaction, while traditional ethical frameworks were abandoned.

Although Western society has emphasized individualism and Asian versions have been more collective, both have centred their efforts on material satisfaction, based often on a rather superficial conception of human potential and needs. Even the communist system, while putting forward social goals, was basically materialistic in orientation. None of these gave any real priority to the wellbeing of the planet and its sustainable environmental management as an essential pre-requisite for our emotional, ethical and spiritual prosperity, as well as our physical survival.[22]

The needed transformation in the economic system requires global public support, which will only come if the proposed solutions are seen as equitable. The combined crises described above call for a sound ethical foundation for human society, including:

+ Justice for all planetary inhabitants
+ Equitable solutions
+ Trust and confidence in the mechanisms
+ Moderation in material development
+ An altruistic cooperative economy
+ Effective wealth redistribution
+ Adequate employment creation.

Agreement on these ethical principles will greatly facilitate the negotiation of solutions at other levels.

Individual responsibility

Redesigning the economic system is primarily an issue for government leaders, businesses and economists. It is not so apparently relevant to actions that each of us as an individual consumer can take. Thus, we need to see the ethical dimensions and consequences of our individual consumer actions.

How is a decision to drive rather than take public transport in a city causally related to the melting of Arctic ice and the plight of the polar bears?

How is it possible to link eating a beefsteak, through the international grain market, to the inability of a poor village woman in south Asia to feed her children?

Why did the rush into biofuels for energy security in the USA lead to food riots in Mexico?

To answer such important questions, we need some basic understanding of the workings of global environmental, economic and social systems. Then we need to ask the ethical questions relevant to our individual contributions to these larger system processes. For example, would you be willing to change your dietary habits to make more grain available for drought-stricken populations? Is it reasonable and just for Europeans to voluntarily accept a reduction in their purchasing power and level of consumption to allow millions in Asia to rise out of poverty? Obviously the answers to such questions are intimately linked to the effectiveness of global systems to deliver food in disaster areas and to ensure that rapid economic growth really benefits the poor – meaning good governance and trustworthiness in the institutions concerned.

Climate change is one area where the consumption patterns of our economy are threatening our future and putting millions of the poor at risk, strengthening the pressure for an economic transition. The United Nations Development Programme, in its *Human Development Report* on the theme of climate change, stated:

> Mitigation of climate change poses real financial, technological and political challenges. But it also asks profound moral and ethical questions of our generation. In the face of clear evidence that inaction will hurt millions of people and consign them to lives of poverty and vulnerability, can we justify inaction? No civilized community adhering to even the most rudimentary ethical standards would answer that question in the affirmative, especially one that lacked neither the technology nor the financial resources to act decisively.[23]

Toward an ethical framework for concerted action

The need to mobilize the world population to respond to these challenges requires new kinds of partnerships across all segments of society. The scientific community can marshal the evidence for income inequality, the poverty trap, over-exploitation of resources, the impacts of pollution and the effects of climate change; but such information by itself does not usually motivate change. Something deeper and more fundamental is needed to build the ethical and emotional commitment necessary for real change. Traditionally, this has been a role of religion; but religion has seldom been seen to have any relevance to the economy. Yet, since rational arguments have been insufficient, a broader approach is obviously necessary.

Religions and faith-based groups are increasingly raising the ethical issues behind the economic challenge, in complement to the scientific arguments. Pope Francis, in his 2015 Encyclical, linked poverty eradication and the environment as ethical issues.[24] Faith-based organizations have a unique reach to grassroots levels all around the world and a capacity to motivate change that can be used to communicate the ethical challenges arising from poverty and climate change, and the need for a common effort to respond. The necessary transition to a just and sustainable economy will require sacrifices from many people and will be more readily accepted when there is an ethical justification or spiritual motivation.

What might an ethical framework for economic transformation towards sustainability based on justice and equity look like? As a start, it will question the dominant materialist society and consumer culture, emphasizing the necessary balance of the material and ethical or spiritual dimensions of human life.[25] By teaching contentment with little and the need to eliminate extremes of wealth and poverty, it would encourage a reconsideration of lifestyles and consumption patterns.

Take from this world only to the measure of your needs and forgo that which exceedeth them.[26]

If we are to address the root drivers of the many imbalances in present world society, we need a more ethical approach to economic development. Economics has ignored the broader context of humanity's social and spiritual existence, resulting in corrosive materialism in the world's more economically advantaged regions and persistent conditions of deprivation among the masses of the world's peoples. Economics should serve people's needs; societies should not be expected to reformulate themselves to fit economic models. The ultimate function of economic systems should be to equip the peoples and institutions of the world with the means to achieve the real purpose of development; that is, the cultivation of the limitless potentialities latent in human consciousness.[27]

We therefore need new economic models that further a dynamic, just and thriving social order, are strongly altruistic and cooperative in nature, provide meaningful employment and help to eradicate poverty in the world.[28] Only such a system will allow us to make the transition to sustainability.

As trustees or stewards of the planet's resources and biodiversity, we must ensure sustainability and equity of resource use into the distant future and consider the environmental consequences of all economic and development activities. We must temper our actions with moderation and humility, value nature in more than economic terms and understand the natural world and its role in humanity's collective development – both materially and spiritually. Sustainable environmental management must come to be seen not as a discretionary commitment mankind can weigh against other competing interests, but rather as a fundamental responsibility that must be shouldered – a pre-requisite for spiritual development as well as the individual's physical survival.[29]

At the social level, we need to focus on the unity of the human race, a unity that is founded on justice and solidarity.[30] Any solution to the challenges facing society must be based on ethical and spiritual principles, inciting individual reflection and community action. For a complex issue such as the economy, where costs and benefits, immediate advantages and long-term risks are so unequally distributed, justice and equity will be essential to achieve any global agreement on action.

> Concern for justice protects the task of defining progress from the temptation to sacrifice the wellbeing of the generality of humankind – and even of the planet itself – to the advantages which technological breakthroughs can make available to privileged minorities . . . Above all, only development programmes that are perceived as meeting their needs and as being just and equitable in objective can hope to engage the commitment of the masses of humanity, upon whom implementation depends.[31]

If we want to meet our enormously demanding collective goals, we need everyone – and every group within society – to put forth relevant human qualities such as honesty, a willingness to work and a spirit of co-operation. We cannot expect everyone to do this unless they can trust that they are protected by standards and assured of benefits that apply equally to all.

As the world moves towards a future global society, the economic challenges are an important force compelling the nations and peoples of the world to give priority to their common interests. Such perspectives of the long-term future of the human race provide a positive focus to counterbalance the concerns for our immediate future.

Accounting

One way both to understand and to manage complex systems is to develop methods of accounting for important stocks and flows in the system. This is how economic systems and businesses are managed. At the simplest level, starting with the amount of money in an account at the beginning, measuring income and expenditures will give a closing balance – which may result in a profit or a loss.

Gross Domestic Product (GDP) is one of the best known economic indicators that is widely used to measure progress in an economy. It is a complex measure of the flow of money within a national economy. The challenge is that the accounting must capture what is really important within the system. However, economic accounting in monetary terms is an overly simplistic way to manage an economy. For example, automobile accidents are good for GDP, since automobile repairs and replacements, as well as medical expenses, increase, but they clearly do not add to human wellbeing.

Using GDP per capita is even less appropriate in ranking the wealth of nations. First, it is an average; therefore, a nation where everyone is reasonably well off and another where a few people are extremely rich and the rest in abject poverty, might rank the same. Second, an increase in per capita income raises wellbeing up to the point where basic needs are met; but higher levels of wealth do not increase happiness or wellbeing and may even cause them to decrease. Other things, like social acceptance, job security or a meaningful life, become more important. Using economic measures to assess and manage larger human systems is inappropriate and misleading. Costs and benefits cannot be calculated only in financial terms.

For example, economists may try to measure the value of a beach by how much money people are willing to pay to go to it; but this only makes sense if everyone has the same ability to pay. A wealthy tourist may pay a lot to access a beautiful beach, but a poor subsistence fisherman – who depends on the beach for his livelihood – may not be able to pay anything.

An accounting system that includes only the flow of money from buying and selling ignores all those things that are not traded in the market and also fails to capture what goes on in the subsistence economy of people who grow their own food and build their own homes and among the poor who barter and exchange in what is often called the informal economy and which is beyond the reach of official statistics.

A further challenge is the over-simplified economic paradigm derived from 19th century physics, which assumes that rational actors, with access to perfect information, use market mechanisms to achieve equilibrium without government interference. In fact, people are far from perfect actors making rational judgements. Information is often hidden or manipulated, the wealthy and powerful are always able to grab more and achieve a monopoly position and the economy is a complex non-linear system driven by dynamic change and innovation far from any equilibrium.[32]

If we want to climb out of this valley, guided by appropriate indicators, we need to acknowledge that there are multiple dimensions in our complex human system, each with its own accounting needs for sustainability. To start with, there are many kinds of capital accounts or stocks: the materials we use and that flow through our system; financial resources; human capital, including labour and social relationships; information and knowledge, including science; and our ethical and spiritual capital of values, principles and ideals. All of these are necessary for civilization and negative trends in the balance of any of them can hinder the others as well.

A full set of both state and flow accounts for a just and sustainable society should therefore include: material accounts (minerals, renewable resources, carbon, food, wastes); energy accounts (energy capture, the energy in fuels and electricity, its flow through the human system and loss to entropy); ecosystem services accounts (water, waste processing, soil maintenance, biodiversity); financial accounts; human capital accounts (labour, education, skills, wisdom, institutions and laws); knowledge/information accounts (science, arts, culture, data); and ethical/spiritual accounts (values, human behaviour, altruism, service to others).

Where the more intangible aspects of human society may be difficult to measure directly, there are scientific ways to assess their presence in human behaviour.[33] Ultimately, this comes down to the power of individuals to do good and contribute to the betterment of society, or to do evil and cause destruction and disintegration. At one extreme there are those, such as the Divine Educators, Buddha, Moses, Christ, Muhammed and Bahá'u'lláh, whose teachings and example have inspired billions over the centuries; at the other are the villains of history, like Hitler, Stalin and many others like them.

What you can do

While you may not be able to change the economy single-handedly, there is already a lot you can do differently in your economic behaviour, at your workplace, or your business. If you are entrepreneurial, you can launch a social enterprise with a principal aim of doing social good while also making a profit. Larger corporations can join the B-corp (benefit corporation) movement, using business as a force for good by placing social and environmental performance, transparency and accountability at the heart of their business plans. Even when you are surrounded by materialism and corruption, you can make the effort to maintain a high standard of conduct in your financial affairs.

Behind these practical considerations are some fundamental principles. Wealth is a means, not an end in itself and its acquisition should not become the goal of one's life. It should be obtained with exertion or through honourable means. Wealth should serve as a means to achieve higher ends, such as meeting one's basic needs, helping one's family to progress and promoting the welfare of society. Furthermore, the end does not serve to justify the means. One should never use improper or unjust means to achieve a goal, however constructive or significant the goal is to oneself or others. The legitimacy of wealth depends on how it is acquired and on how it is expended. It is praiseworthy when it is earned through one's own efforts – in commerce, agriculture, crafts and industry – if it serves to enrich everyone and if it is used for benevolent purposes, the promotion of knowledge and the common good.[34]

The acquisition of wealth should be governed by the requirements of justice. For example, an employer and employee are bound by the laws and conventions that regulate their work, and are expected to carry out their responsibilities with honesty and integrity. At a deeper level, justice should determine the relationship between the minimum wage and the cost of living. Workers contribute to a company's success and are entitled to a fair share of the profits. Some of the unjust practices in today's economy are the wide margin – often unjustifiable – between the production costs of certain goods and the price at which they are sold, the exploitation of others, the monopolization and manipulation of markets and the production of goods that promote violence and immorality.[35] Everyone should be careful not to be drawn into situations involving these practices.

The Sustainable Development Goals

The year 2015 was an important one for the world community, with agreement on a new 2030 Agenda for Sustainable Development and the Paris Agreement to address climate change, among others. The UN Secretary-General called for a fundamental transformation in society and the economy. The Sustainable Development Goals (SDGs) at the heart of the new agenda define a paradigm shift for people and the planet that is inclusive and people-centred, leaving no one behind. The new agenda integrates the economic, social and environmental dimensions of development in a spirit of solidarity, cooperation and mutual accountability, with the participation of governments and all stakeholders.[36]

The world leaders assembled at the UN General Assembly Summit in September 2015 stated the following in their outcome document:

This Agenda is a plan of action for people, planet and prosperity. It also seeks to strengthen universal peace in larger freedom. We recognise that eradicating poverty in all its forms and dimensions, including extreme poverty, is the greatest global challenge and an indispensable requirement for sustainable development.

We are resolved to free the human race from the tyranny of poverty and want and to heal and secure our planet. We are determined to take the bold and transformative steps which are urgently needed to shift the world onto a sustainable and resilient path. As we embark on this collective journey, we pledge that no one will be left behind.

It is 'We the People' who are embarking today on the road to 2030. Our journey will involve Governments as well as Parliaments, the UN system and other international institutions, local authorities, indigenous peoples, civil society, business and the private sector, the scientific and academic community – and all people. . . It is an Agenda of the people, by the people and for the people – and this, we believe, will ensure its success.[37]

The 2030 Agenda can be summarized as follows:

People

We are determined to end poverty and hunger, in all their forms and dimensions and to ensure that all human beings can fulfil their potential in dignity and equality and in a healthy environment.

Planet

We are determined to protect the planet from degradation, including through sustainable consumption and production, sustainably managing its natural resources and taking urgent action on climate change so that it can support the needs of the present and future generations.

Prosperity

We are determined to ensure that all human beings can enjoy prosperous and fulfilling lives and that economic, social and technological progress occurs in harmony with nature.

Peace

We are determined to foster peaceful, just and inclusive societies which are free from fear and violence. There can be no sustainable development without peace and no peace without sustainable development.

Partnership

We are determined to mobilize the means required to implement this Agenda through a revitalized Global Partnership for Sustainable Development, based on a spirit of strengthened global solidarity, focused in particular on the needs of the poorest and most vulnerable and with the participation of all countries, all stakeholders and all people.[38]

The 2030 Agenda is more than just a declaration without

substance. It includes 17 Sustainable Development Goals (SDGs), which are action-oriented, global in nature and universally applicable. Supporting the 17 goals are 169 quantified targets to be achieved by 2030 and global indicators have been identified to measure progress towards the targets. Governments are now

The 17 Sustainable Development Goals[39]

1. End poverty in all its forms everywhere.
2. End hunger, achieve food security and improved nutrition and promote sustainable agriculture.
3. Ensure healthy lives and promote wellbeing for all at all ages.
4. Ensure inclusive and equitable quality education and promote life-long learning opportunities for all.
5. Achieve gender equality and empower all women and girls.
6. Ensure availability and sustainable management of water and sanitation for all.
7. Ensure access to affordable, reliable, sustainable and modern energy for all.
8. Promote sustained, inclusive and sustainable economic growth, full and productive employment and decent work for all.
9. Build resilient infrastructure, promote inclusive and sustainable industrialization and foster innovation.
10. Reduce inequality within and among countries.
11. Make cities and human settlements inclusive, safe, resilient and sustainable.
12. Ensure sustainable consumption and production patterns.
13. Take urgent action to combat climate change and its impacts.
14. Conserve and sustainably use the oceans, seas and marine resources for sustainable development.
15. Protect, restore and promote sustainable use of terrestrial ecosystems, sustainably manage forests, combat desertification, halt and reverse land degradation, and halt biodiversity loss.
16. Promote peaceful and inclusive societies for sustainable development, provide access to justice for all and build effective, accountable and inclusive institutions at all levels.
17. Strengthen the means of implementation and revitalize the global partnership for sustainable development.

expected to adapt the goals and targets to their own national priorities and realities, to determine their share of responsibility for the global goals and to report regularly on their progress.

These are aspirational goals that we can see are relevant and motivating. The fact that they have been approved by world leaders and the United Nations gives them added weight. And since they need to be achieved by 2030, we all can have an important role in making them come true and in holding governments to account for their commitments. However, governments are not very good at keeping their promises; we also need a transformation in people and in the institutions of society.

These goals are also relevant to transformation at the community level through social action, as we have seen previously. For example, in the following box, the goals and targets have been rewritten in ways that we, as individuals, or our social group or local community, can take responsibility for and put them into action where we live and work.

Sustainable Development Goals for individuals and communities

Goal 1: No poverty

Contribute to local efforts to eliminate poverty in our community

Goal 2: Zero hunger

Support community efforts to ensure everyone access to safe, nutritious and sufficient food all year round
Encourage and support local small-scale food producers
Support sustainable food production systems that improve land and soil quality

Goal 3: Good health and wellbeing

Choose a healthy lifestyle for ourselves and our families
Avoid narcotic drugs and harmful use of alcohol
Drive safely
Plan our family size
Avoid using hazardous chemicals and try not to live in polluted areas

---

Goal 4: Quality education

Get the best education possible and educate our children

Give our small children pre-primary education

Help others to get skills for employment, decent jobs and entrepreneurship

Encourage education for girls and vulnerable populations

Educate ourselves, our families and community about sustainable development and sustainable lifestyles, human rights, gender equality, peace and non-violence, global citizenship and appreciation of cultural diversity

Goal 5: Gender equality

Avoid discriminating against women and girls

Shun violence against women and girls

Share responsibility within our household and family

Encourage women's participation in leadership and decision making

Support women's equal rights

Promote the empowerment of women with technology

Goal 6: Clean water and sanitation

Encourage safe drinking water and sanitation, practice good hygiene

Avoid polluting water

Use water efficiently

Contribute to improving water and sanitation in our community

Goal 7: Affordable and clean energy

Prefer renewable energy sources

Use energy efficiently

Goal 8: Decent work and economic growth

Consider a career in a sustainable productive activity involving creativity and innovation

See our work and that of others as a service to the community

Help young people to find training and employment

Encourage all workers' – including migrants – rights to a safe and secure working environment

Goal 9: Industry, innovation and infrastructure

Work to improve our local community infrastructure
Look for ways to make our workplaces more resource efficient and sustainable
Learn to use information and communications technologies and help others

Goal 10: Reduced inequalities

Participate in the life of our community and empower others irrespective of age, sex, disability, race, ethnicity, origin, religion, economics or any other status
Support equal opportunities for everyone in the community
Be welcoming to migrants

Goal 11: Sustainable cities and communities

Choose our housing to be safe and sustainable
Use sustainable forms of transport
Participate in the sustainability planning of our local community
Protect our local cultural and natural heritage
Reduce our vulnerability to disasters
Contribute to community gardens and green spaces

Goal 12: Responsible consumption and production

Consider sustainable natural resource use in our purchases
Stop wasting food
Reduce our use and release of chemicals
Reduce our wastes through prevention, reduction, recycling and reuse
Inform ourselves and help to educate others about sustainable development and lifestyles in harmony with nature

Goal 13: Climate action

Educate ourselves and others about climate change mitigation, adaptation, impact reduction and early warning

Goal 14: Life below water

Reduce our use of plastics and dispose of them responsibly
If we live near the coast, support coastal (river, lake and lagoon) protection
Avoid releasing chemicals and other toxic products into the water

Goal 15: Life on land

Support the conservation and sustainable use of terrestrial and fresh-water ecosystems, especially forests, wetlands, mountains and drylands
Use paper, wood and charcoal from sustainable forestry
Protect local natural habitats and biodiversity

Goal 16: Peace, justice and strong institutions

Avoid all violence
Protect children from abuse
Fight local corruption
Demand accountability and transparency from our local institutions
Participate in local decision making
Avoid all discrimination in our community

Goal 17: Partnerships for the goals

Contribute time and resources to local sustainability efforts
Invent, adopt and share environmentally sound technologies
Join in local partnerships for sustainability

In the next valleys we shall explore what else is required to make the just society defined by these goals a reality.

5

The Valley of Individual Discovery and Development

You must have realized – as you climbed through the valleys and up the mountains of environmental crises and sustainability, social illness and justice and economic crisis and transformation – that so much depends on the behaviour of each individual human being and even more so for those who are in positions of power and authority. The institutions of society, no matter how well conceived, will disfunction if the people within them are self-serving, dishonest or corrupt. This is what traps people in these valleys.

On the other hand, even poorly-designed institutions can perform well if those within them have high ethical standards and are motivated to be of service to others. It is not enough to transform the structures and processes of society in any field if people are not also changed. While we may not be able to do much about others around us, we can always start with ourselves. And as we learn about self-transformation, we can both become an example for others to follow and accompany others on their own life journeys.

This fifth valley is the valley of individual discovery and development, especially your own. Where the threats and challenges of previous valleys were all around you, in this valley they are within you. The valley bottom is like a maze enclosed in walls and high hedges, in which you wander through endlessly branching paths. Mostly you are alone in this valley, but you often come upon mirrors that reflect your own self. Occasionally, at a branch in the path, you may meet another wandering soul who may share something about how others see you, before turning some other way. Only self knowledge will enable you eventually to find a way out of this valley and towards the heights beyond.

While you want to be one of the many well-intentioned individuals who are working to improve the society around you, you may often feel that you face insurmountable obstacles. Your hopes may flounder if you share the erroneous assumptions about human nature – that we are incorrigibly selfish and aggressive and incapable of creating a just social system. These assumptions so permeate the structures and traditions of present-day living as to be considered established fact. They appear to make no allowance for the extraordinary reservoir of spiritual potential available to us and to any illumined soul who draws upon it. Instead, they rely for justification on humanity's failings, examples of which daily reinforce a common sense of despair. These false premises obscure the fundamental truth – that the state of the world reflects a distortion of the human spirit, not its essential nature.[1]

So how do we transform ourselves and develop the spiritual capacities to contribute to a process of societal change?

Human nature

The first obstacle in this valley is our understanding of our own human nature and purpose. How often have we heard: 'You can't change human nature'? Yet, this all depends on how we define human nature. While there is a genetic component to intelligence and personality, the strongest influence is education, which can override many genetic predispositions. And then, is human nature only our lower animal side – most obvious in the immature stages when we were an infant and child – or is it the full potential of what we can become, physically, socially, emotionally, intellectually and spiritually?

Perhaps this idea that we cannot change is really an excuse for not wanting to make the effort to change, looking for the easy way rather than struggling to master our lower desires in an effort to refine our character and to contribute to society in some constructive way.

Take some time to consider what really is human nature or human potential. We, in fact, have three realities: physical, intellectual or rational and spiritual. Our physical reality is our body, which developed as an embryo in the womb of our mother; was born as a helpless infant; gradually completed its brain development and mastery of motor skills to the point where we could walk and talk; learned by observation and imitation, then by language and reading; acquired social skills; and passed or is passing through adolescence to increase strength, become sexually mature and end up as a fully-functioning adult.

However, many of our physical qualities peak in early adulthood and, from then on, it is physically all down hill, through menopause and ageing, until finally one disfunction or another leads to our death. Our physical perfection is a very transient thing and it is obviously not worth putting all our hopes on it. The real significance of physical qualities is that they enable us to advance at other levels of reality.

Our intellectual reality is less tangible. The knowledge and skills we acquire may help us in our profession – whether as a farmer, artist or professor. But while knowledge can accumulate in our minds for decades and mellow into wisdom, our intellectual powers also go into slow decline from early adulthood. As we age, our memory also may start to fail, first in retaining new memories, then in retaining knowledge previously acquired.

Late in the ageing process, we may even suffer from dementia, lose autonomy and regress to a kind of child-like dependence on others. Our intellectual reality may allow us to make important contributions to our family, community and all humanity, but it is no more permanent than our physical reality.

Our spirituality or spiritual reality is even less tangible than our intellect, since its most evident characteristic is love and it is created by developing virtues and abstract qualities. All cultures acknowledge it in one form or another. All religions proclaim its existence and describe its cultivation as our real purpose in life. In fact, it seems to be what connects us to other planes of existence beyond this physical universe (as we shall explore in the next valley) and, being intangible, its existence is not terminated by the death and disintegration of our body. No experience in this life would allow us to imagine what the progression of our individual spiritual reality is like, but we can rely on the superior knowledge of the great spiritual teachers and founders of religions, all of whom certify its reality, just as we rely on scientists to confirm the reality of quantum physics or the neurobiological processes in the brain.

While memory loss, ageing and death may seem like a tragedy with respect to our physical and intellectual realities, they in fact find their logical purpose in helping us to develop our spiritual reality. Detachment is an important driver of spiritual growth. The acorn must be detached from its reality as a seed and sacrifice itself to grow into an oak tree. If we cling to our childhood, we cannot mature properly as an adult. In the same way, our pride at becoming rich, being a champion athlete, or winning a Nobel Prize for intellectual achievement, could be a barrier to our further spiritual development and our relationships with other people. Acknowledging that these achievements are ephemeral can strengthen our humility and discourage us from seeing ourselves as being better than other people. Accepting that you have a spiritual reality to cultivate opens the path to your upward journey out of this valley.

Egoism and altruism

The maze of the ego tries very hard to keep you forever wandering in circles within it. We start as a child by building our own identity and are naturally self-centred. Growing up provides many opportunities to become more other-centred, especially when we bear and raise children. Our body, our physical reality, is temporary and its death is inevitable. Our spiritual reality takes form in this life and is the ultimate purpose of life itself. In fact, the human lifespan lends itself to this process: we acquire physical strength, intellectual and social skills and knowledge in our youth; make use of them in adulthood; and then detach ourselves from them as we lose them again in old age, so that ultimately all that remains is our spirit – even before death. This is the way to escape from the maze.

In considering the three levels of human reality, in our physical reality we can see the ego expressed as hedonism. Physical pleasure is what counts, whether through sex, alcohol and drugs, or food and comfort. At the intellectual level, the ego is expressed in pride at what we know and are discovering and the belief – common among many scientists – that we have the capacity to ultimately know everything. From this materialist perspective, reality is what we are capable of knowing and proving scientifically. Reality stops there and everything else is superstition or imagination.

However, if we acknowledge that there is an ultimate, absolute reality and that our purpose is to approach it through human consciousness and spirituality, this changes everything and opens a whole new potential. We can be content with little and freed from all inordinate desire at the physical level – meeting our needs, but then turning to more important things in life. We can have the humility to acknowledge that our human mind, language and intellectual tools are limited by our own experiences and, while science can always progress, we can never know everything. Science can in fact free us from some of the struggles of this material existence for higher purposes. We can then turn our attention to our true purpose in life to develop the limitless potential in our human consciousness and spirituality.

It is in the nature of all human beings to be both egoistic and altruistic and life can be seen as a struggle between these two contradictory tendencies. Self-centredness is the fallback position, the original infantile state. Cultivating other-centredness requires education and effort, as we grow from a child into a mature adult.

Why should we make the effort, when thinking of ourselves first is more immediately satisfying? Because this is what both allows us as individuals to develop our potential and facilitates the complex social interactions upon which a successful civilization is built. Most of the problems in society today can be traced to neglect of this fundamental dimension of human potential. This valley, with its winding maze, is the most difficult to cross and the climb out of it is both the most arduous and the most rewarding.

At present, the focus on material civilization cultivates the lowest dimensions of human nature. Many leaders in all fields are prisoners of their own egos, power-seeking and greedy and their short-term successes are ultimately destructive, leading to war, terrorism, organized crime and corruption. Consider the enormous impacts of ego, pride and selfishness on society, whether it be dictators and tyrants ruling countries, generals in warfare, corporate leaders maximizing profits above all else, religious leaders basking in their glory, or the rich vying to flaunt their wealth. Power just magnifies the negative effects of individuals with strong egos.

Transformation must start in childhood, when education to good character takes place – largely in the family and community. Children who have responsibilities and contribute to the life of the family, through chores or other forms of service, have a greater sense of confidence and self-worth. There is nothing wrong with self-knowledge and an appreciation of the positive qualities we have acquired, just as it is important to be aware of our weaknesses and failings and the need to overcome them.

In pre-adolescence, parental control weakens and we lay the foundation of values and behaviours that set the course of our life. We ask questions such as: Why am I here? What is my purpose in life? This is a time of idealism, when everything seems possible. Our life can either veer towards self-indulgence and hedonism, or we can learn the deeper satisfaction of being of service to others. This time is particularly challenging for youth in the modern world because our consumer society cultivates passive consumers of immediate pleasures, while playing on animal desires and feeding the ego. Learning the values of altruism is really like swimming against the current.

Mastering our ego empowers us to evolve towards fulfilling our higher human purpose, to refine our character and discover new qualities we did not know we possessed. When we are no longer enslaved by our ego, we have the free will to create new possibilities. Forgetting ourselves and becoming detached from the material things of this world frees our potential to evolve individually towards endless perfections.

Think of the multiplier effect – of having many individuals all striving to become altruistic, selfless and full of love, desiring to be of service to others. This is what will enable us to evolve collectively, to extend social relationships, to build communities and institutions and to advance civilization. It will allow human society to achieve higher orders of integration, cooperation and reciprocity, both geographically and over time. Now that science and technology have removed all the physical barriers to global integration, the process of building a world civilization is beginning and we can all be part of it.

Optimism and pessimism

In this valley of internal struggles, it will also help to understand where we stand on the spectrum of optimism and pessimism. The optimist is sure there is a way out of the maze while the pessimist is convinced that there is no exit. Psychological research shows that these are two common attitudes to life.[2] Life is never easy, but we can respond to the difficulties in positive or negative ways. An optimist sees a better future ahead that will help us overcome our problems. A pessimist feels guilty about bad moments and this spreads to become a general view about life. While genetic predisposition may account for a quarter of the tendency to one or the other, with a stronger fear response associated with more pessimistic and conservative world views,[3] it is largely a matter of choice, education and the accidents of life. Research shows that a secure maternal attachment during infancy can help create self-confidence,[4] but that is not essential.

Some educational systems emphasize being critical of everything and where competition to be better means putting down others. The result can be an entire society with a tendency to see the bad side of things. Such attitudes can be self-reinforcing and pessimists tend to withdraw and isolate themselves from social contact or select the media and seek out others who confirm their worst suspicions.

People differ in their biologically determined negativity bias. We subconsciously respond more and pay greater attention to negative rather than positive events and individuals vary in their degree of response to negative stimuli, with some reacting almost equally to positive and negative and others much more strongly to negative. One study has shown that political conservatives react more fearfully than liberals to threatening images and see the world as a dangerous place. They have stronger reactions of disgust to bad odours and morally-suspect behaviours. Those with a strong negativity-bias favour protective policies are suspicious of new approaches and people who are different and prefer certainty, tradition and security. They seek in-group safety and react against immigration and other perceived threats. Individual political orientations such as attraction to populism are deeply connected to biological forces that are usually beyond personal control and are grounded in emotions.[5] Education and life experience can help to overcome negative emotions and encourage positive emotions.

People with different political beliefs come to inhabit different realities. It is our moral positions that determine what facts we accept. We may jump quickly to moral conclusions and then come up with reasons to justify our decisions. If the facts appear to contradict a moral position, we may dispute the facts or come up with alternatives that support our beliefs. What feels right to believe is shaped by the culture we grew up in, with many of our fundamental beliefs formed in childhood. We are social beings and learn beliefs from the people we are closest to. Our beliefs are part of our cultural identity and this often has greater determination on our beliefs than do facts.

Optimism can also be contagious. If we feel down, it can help to associate with optimists who see the good side of things, who remember the good times and who do not dwell on the unpleasant things of life. Optimists understand that difficult situations result from specific conditions that are limited in time. The intelligent optimist is in touch with reality, neither hiding behind a forced optimism nor lost in an unrealistic ideal. They know themselves, are honest about what they are capable of without guilt or shame and look for the positive side of things.

The classic story illustrating this tells of Jesus walking with some friends who come across a dead dog. When the others who are with Him comment on the stench and ugliness of the dog, Jesus responds that it has beautiful teeth. Optimists have lots of advantages: they are more popular, as others seek them out to be uplifted; they are more curious to explore and are free to imagine, dream, play and have fun, as well as to let go; they find pleasure and inspiration in music, reading, art, beauty and contact with nature; and they have a greater tendency to be happy.[6]

While there are many good reasons to be pessimistic today and the first part of this journey is full of these reasons, why not choose to be optimistic? It is the first step in empowering yourself to be an agent of change. It is the rational foundation of hope. It is the best weapon in the combat for a better future. And it is the fastest way to lift yourself out of this valley.

The individual and society

If we consider what makes a civilization progress, we can see a number of factors at work. Perhaps the most important is social cohesion.[7] People need to be motivated to work together and even to sacrifice for each other (as we saw in the valley of social justice). Each individual needs to accept the primary importance of the group.

This should then lead to an effective system of organization and governance in society, as well as to intellectual, scientific and technological advances that permit greater utilization of resources and technological progress. Note that these are largely internal to the society, although they may also build on previous civilizations, as, for example, when Europe in the Renaissance turned to Greek, Roman and Arabic sources and borrowed ideas from other contemporaries.

The same is true of us as individuals. Our progress and the

realization of our potential as human beings is largely the result of our own efforts, although we also require values and knowledge that are acquired through education. Knowledge is useless if it is not put into action and values are pious hopes if we do not live them.

Fortunately, change is possible, but it is not easy. It is challenging because it is difficult to imagine what change in the future would look like. Perhaps the most important factor is our own belief that we can change. If we start with this, change *can* happen. Change also requires courage. We need to be brave to face our inner shortcomings or external threats, not to give up or yield to pressures to conform and to continue even when the going gets tough. Courage can be learned. If our moral foundation is strong enough, we can resist anything, even if it means giving up our life for our beliefs, as many examples in history attest.

This leads each of us to one of the most critical questions that we will face: *What am I going to do with my own life?* Just as a scientific discovery can be put to good or bad uses, so can our human potential be turned to good or bad ends, or be fulfilled with good or bad means. The choice is up to us, no one else. We can become a greedy banker building a fortune at the expense of others, an employer exploiting poor labour, a corrupt politician or a drug trafficker. Or we can make important scientific discoveries, organize humanitarian assistance, devote ourselves to teaching young people or create beautiful works of art. And even when the end is beneficial for society, we can be driven by pride, ego or a desire for fame and recognition, stooping to anything to achieve our end. Notice the difference between these different kinds of successes: a promotion of self and disregard for the suffering of others, or a humble desire to be of service to humanity.

The biggest challenge you face in life and in this valley is yourself. We are born with the potential for both good and evil;

with evil being the absence of good as darkness is the absence of light. As a child, it is natural for us to think of ourselves first and to form an ego. But, if we want to be an adult who is successful and beneficial to society and ultimately to ourselves, we must then learn to master our inner egotistical side and to turn outwards. This is never easy and it is a struggle that will continue for all of our life – for the selfish side of ourselves is never totally defeated.

This is a universal theme in all civilizations, from Greek mythology and tragedies to modern literature. It is also at the heart of all religious traditions and, in fact, religion could be described as the dimension of civilization that is specialized to address this part of us. Saints and sinners, heaven and hell, reincarnation, nirvana, salvation, the right way – these are all ways of giving form to our basic struggle with ourselves and the options before us. And all are accompanied by tools that can help us on our journey.

Those who say that you cannot change human nature generally believe that humans are fundamentally selfish and aggressive. They are not entirely wrong, but this theory is incomplete. We also have the potential to be altruistic and peace-loving; but we need education to give us this vision of our higher human purpose and the strength of character to put it into action.

Why is this so important? Why does it matter if we are a teacher or a murderer? Apart from the individual outcome (i.e. thankful students or prison), it is really in the interest of society and of human advancement as a species and as a builder of civilizations, that we are constructive rather than destructive. If we look at the problems in society today, they can almost all be traced back to selfish behaviours and a lack of ethics or values. Societies advance when there is trust, cooperation and solidarity and decline when they are dominated by corruption, competition and exploitation.

The evolution of any species is dependent on its survival and on its success in adapting to changing conditions, both as individuals and as a species. This is how nature works. For the human species, our evolution today takes place largely at a social level. We can only survive as individuals within a functioning family, community and society. In a world that has become united through science and technology, we now have to evolve rapidly to make the transition to social and spiritual unity in a global civilization, or risk crises that could wipe out a major part of humanity.

This presents each of us with our own individual challenge: do we want to be part of the problem or part of the solution? Do we simply let ourselves be swept along in the current, behaving and consuming like everyone else, or are we ready to make the effort to swim against the tide and become an example for others to follow? It is hard to be a pioneer and to risk ostracism, or worse. The struggle to master ourselves is hard enough; so much more so when those around us do not believe in the need for it. This is the choice we face today if we want to build a better world.

There are many idealistic movements and people working for good causes, but too often they fail, or their success is tarnished because they suffer from the same individual human failings as the rest of society. Our individual transformation is at the heart of the matter and everything else follows from that. Your commitment to that transformation and your continuing effort throughout your life will give you the strength to climb up the path out of this valley.

Rationality and belief

Returning to our three human realities – physical, rational and spiritual – it is important to understand their complementarity, their roles in our life and how they grow and develop. This can also help us to understand the relationship between the two great knowledge systems of science and religion.

The scientific knowledge system is evidence-based, with facts that can be tested through observation and experimentation, building an understanding of our physical reality that is continually advanced and perfected. This does not mean that scientific truth is anywhere near absolute. Concepts that seem firmly established may suddenly be overthrown by new theories, new evidence and new mechanisms to explain existing facts.

A few decades ago, the continents were considered stable on the surface of the planet, and observations that some continents looked like pieces of a puzzle that might fit together were considered fortuitous. A few years later, data on the young age of ocean bottoms and a new understanding of the mantle in the planet's interior proved that continents in fact drifted across the planet's surface, breaking up and colliding.

The science of physics has similarly undergone revolutions in the last century. Even scientists trying to be purely rational may observe what they expect to observe and find ways to explain away data that does not fit. Peer review, which is intended to ensure that proper scientific rigour is observed, may also reject research that is too far outside the established paradigm, but which may subsequently prove to be an important step forward.

Religion is considered one form of belief, but there are many others and it turns out that belief is as fundamental to human life as rational thought. Recent research is deepening scientific understanding of the importance of belief and raises fundamental questions about how to consider its role in our own lives and in society.[8]

Beliefs define how we see the world and act within it. They are what make us human. They tell us what is right or good and thus how to behave towards others and the natural world, but they are difficult to define. Our brain tries to extract meaning from all its inputs. Knowing something is true is different from believing it to be true. Knowledge is objective and belief is subjective. Brain research shows that we unthinkingly accept what we learn in childhood as true and we have to make an effort to doubt and reject it. Disbelief requires much more brain activity in regions associated with deliberation and decision making, as well as with emotions of pain and disgust. Belief involves both reasoning and emotion.[9]

Belief may lead us to accept as true things that are unscientific or irrational, such as conspiracy theories, the paranormal, superstitions and magical thinking. We can even believe in contradictory things. Beliefs are what lie behind feelings of racial or national superiority and make it easy to be polarized into 'us' versus 'them' – which is at the root of many social conflicts. If a belief is challenged, we can become defensive, rigid in our resistance and reject the facts and the experts who delivered them.[10] Here, our challenge is to become aware of any contradictions and to work for coherence between our rational self and our beliefs.

More about beliefs

In the normal process of belief formation, we combine incoming information with unconscious reflection on that information until it feels right – and a belief is formed. People are surprisingly susceptible to strange beliefs, especially beliefs that we cannot easily verify with our senses. Most people accept as true things that are unscientific, if not delusional. Half of adults in the United States endorse at least one conspiracy theory and 90 per cent of adults in the United Kingdom hold at least one delusional belief – such as not being in control of some of their actions, or that people say or do things that communicate secret messages intended for them alone. The feeling of 'rightness' is also fallible and is based on our evolved psychology, personal biological differences and the society around us.[11]

Our evolved psychology is closely linked to religion as a belief system, with religious belief remarkably similar across all religions, including some supernatural agency, life after death, moral directives and answers to existential questions. Our brain is primed to see agency and purpose everywhere as it searches for meaning. For the cognitive by-product theory of religion, this is merely an attempt to give meaning to otherwise random events. Where religious claims are frequently encountered in early childhood, they are unquestioningly accepted and rooted deeply in our cognitive architecture and feelings of rightness. The same process can make us susceptible to many irrational beliefs – from the paranormal and supernatural to conspiracy theories, superstitions, extremism and magical thinking. It also supports the dualistic belief that our mind and body are separate entities and the belief that the group we belong to is superior to others.[12]

Social psychology has also shown that we can believe in contradictory things without being aware of their incoherence. A good example is attitudes towards migrants or foreigners. Our society places value on national identity and culture, which we do not want to see eroded. On the other hand, we may have a tradition of tolerance and openness to others, especially if refugees have escaped from war or persecution. Psychologically, we can believe both at the same time – by hiding our prejudices as implicit or contextual, so that we retain a self image as one free of prejudice, while putting the blame on the other person, such as by saying 'I'm not racist, but …' Research shows that the more we know migrants personally, the less likely we are prejudiced against them, while those who have simply been told about conflicts with migrants are more vulnerable to anti-migrant propaganda. If the social discourse is polarized into 'us' versus 'them', or the 'other' is depicted as inferior or undemocratic, we more easily accept violence against 'them'.[13]

A similar process works with respect to scientific facts which may be in contradiction with our beliefs or behaviours, such as smoking. We can become defensive and reject the facts and the experts that deliver them. We may be troubled because we cannot tell the experts that they are wrong, so we become more rigid in our resistance, refuse to think about the problem, or become evasive.[14]

Individual beliefs

By the time we reach adulthood, we should have a relatively coherent and resilient set of beliefs for the rest of our lives, which can vary along five independent dimensions of what are considered to be worthy sources of value and goodness in life.[15] These are:

- **Traditional religiousness** – the level of belief in mainstream theological systems, such as Christianity and Islam.
- **Subjective spirituality** – the level of belief in non-material phenomena, such as spirits, astrology and the paranormal.
- **Unmitigated self-interest** – the belief in the idea that hedonism is a source of value and goodness in life.
- **Communal rationalism** – the belief in the importance of common institutions and the exercise of reason.
- **Inequality aversion** – the level of tolerance of inequality in society, a proxy of the traditional left-right political split.

Our beliefs thus have little to do with conscious rational choices and are highly resistant to change. We can go to great lengths to reject something that contradicts our position, or seek out further information to confirm what we already believe. However, we can and do change our minds, usually not rationally but in response to compelling moral arguments and will reshape the facts to fit with our new beliefs. These beliefs are the deep roots of many political, religious and social troubles, but are largely invisible to us.[16]

For the rational materialist, it is unsettling to discover that people have little conscious control over their beliefs – which are built on intuition, biases and gut instincts. Even scientists are influenced by their beliefs about what is important, what they might find and what their findings mean. Belief is a potent force in human affairs and the foundation of civilization. For most scientists, belief without supporting evidence or argument should be rejected as a basis for politics or policy.[17] The same argument is used to reject religion. Yet science is also limited. It is a reliable basis for understanding our physical reality and can provide tools for use with our intellectual reality. But, much of our spiritual reality cannot be subjected to any kind of scientific measurement. Does this mean that it does not exist or should not have an influence on our lives? We cannot measure love by weight or energy content, but does it therefore not exist?

Perhaps, rather than fighting against belief, rejecting it or trying to reduce it to what science can understand, we should ask how we can turn belief from something negative and that is the cause of irrational behaviour, prejudice and conflict, into a force for unity and peace. Perhaps good beliefs that contribute to the advancement of civilization can be cultivated. This is one of the fundamental themes all along this journey.

While the cognitive by-product theory of religion reduces it to an attempt by the brain to find meaning where there is none, we can also ask what constructive role religion has frequently played for individuals, communities and whole civilizations. Individuals and institutions all go through cycles of growth, maturity and decline or decadence. Do we focus only on the latter, observing the admitted disfunctions of old religions, or see what can be gained by considering how religion can form a positive and scientifically justified belief system adopted to the needs of today?

One issue in dealing with belief is to ask if there is an equivalent to empirical evidence in science as a measure of truthfulness or rightness of belief. Is there a touchstone of truth with which beliefs can be tested? Is there an authority other than scientific authority that can guide the selection of beliefs that are constructive rather than destructive of human wellbeing and advancement? In particular, religion has usually claimed some kind of divine authority for its basic teachings and beliefs, but this has not prevented the emergence of conflicting dogmas and disputes. This is one challenge awaiting you in the valley to come.

All of the above demonstrates the human need for an Educator. We are not born with a genetically determined set of values, or the instinctive behaviour of animals. We have to learn our values in ways that protect us from some of the irrationality in our brain functions. Scientific education teaches us about the world around us and the rules we should respect to survive and prosper physically. But we equally need some form of spiritual education to give us a sense of purpose in life and the values and ethical rules necessary for a healthy society. One of the original purposes of religion has been to provide Divine Educators for humanity at various stages in its development. However, too many tragedies have been inflicted on humanity through the wrong forms of spiritual education, so it is important to avoid indoctrination and to respect our individual responsibility to investigate the truth and our freedom to choose our own beliefs and values.

Your beliefs and values will be the armour that will protect you on your continuing journey upwards. Are they fit for purpose? In your crossing of this valley, have you identified a need to re-examine some of your beliefs and to bring them into coherence with the goals you have set for yourself on this voyage of self discovery?

Cultivating your spiritual reality

As you work your way through the mazes up this valley, in addition to the armour of your beliefs and values there are some weapons you can use to defend yourself against your ego and tools that will help you to find and cultivate your spiritual reality. Fortunately, universal human problems also have some universal remedies and thousands of years' experience with this has been captured in the great religious and spiritual traditions of the world. By comparing across traditions and finding what is common among them, we can separate what is universal from the rituals and dogmas with which religions tend to become encumbered over time. Some of these, like belief in some absolute truth or unknowable essence, will be explored in the next valley.

As you emerge from this valley, you will find houses of worship perched on the hillsides – quiet places where you can explore the power of prayer and calm and peaceful sites for meditation. To learn detachment, you can walk the paths of fasting and there are libraries, the internet and e-books giving you access to all the world's spiritual knowledge and wisdom – including the holy scriptures of all religions – for study and reflection. These tools described below will provide some more practical exercises to refocus your attention on what is most important for your own development and fulfilment as you climb towards the peaks of self knowledge.

Prayer

It is in the nature of humans as social beings to want to talk about things and to express our feelings. We need to verbalize to put our thoughts and feelings into words. Prayer is speaking to God, or whatever we call the Unknowable Essence, and expressing our love towards this Divine Source. Prayer may also be asking something from God, be it love, assistance or forgiveness. Although God is above and knowledge and understanding and presumably already knows everything in our minds and hearts, it can help us to put words to our feelings.

Some religions have revealed prayers or scriptures that can capture these feelings better than we can ourselves and that help to educate our spiritual natures. This is why prayer is an almost universal part of religion and belief; turning towards the object and source of our belief and requesting or beseeching something. Since cultivating our spiritual reality requires detaching ourselves from our material reality, ego and superficial pleasures, prayer becomes an essential part of that process of turning outward, re-directing our positive feelings of love and our desire for comfort and assistance, to this external focus. If we are able to purify this prayer from idle fancies and vain imaginings, selfish thoughts and material desires, it will have its impact and lead us in the right direction.

While we can make up our own prayers, we risk projecting a selfish perspective. If we have the benefit of revealed prayers, designed for our education, their impact will be more profound since they can put our personal issues and concerns into a new and more spiritual perspective. Daily prayer is important to keep reminding us of our spiritual priorities. If we are not constantly advancing, we risk slipping backwards.

Meditation

Meditation is another tool for the emptying of self and reflecting deeply on spiritual things, but without necessarily putting them in words. Different religious traditions have their own approaches to meditation, but they all contribute in their diversity to spiritual development. They may involve deep thought and reflection on a prayer or spiritual theme, or a calming and even emptying of the mind, focusing on something regular and essential like breathing.

Meditation can help us to listen to our inner voice, put

things into perspective and lead us to a decision or course of action. It can be a prayer that rises above words and letters, syllables and sounds. It is a quiet time that can be an excellent antidote to the stress and tensions of the modern world. We can all find a time and a form of meditation that suits us.

Fasting

The great religious traditions generally include some form of physical discipline or restriction to help us to recognize the priority we should give to our spiritual development over our material desires. Fasting, or refraining from eating and often drinking, is one of the most common – whether in the Christian Lent, the Muslim Ramadan or the Bahá'í Fast. Those who fast generally find that breaking the routine of eating and drinking at will helps to acquire spiritual qualities, such as patience and detachment.

Voluntarily giving up the priorities and pleasures of the body is a way of acknowledging that spiritual development is ultimately more important. Self discipline learned in this way can carry over in many other beneficial ways, especially in a society that gives such importance to immediate hedonistic pleasures.

Daily study

As with prayer, we need to work on our spiritual growth every day of our lives, or we too easily become forgetful and slip backwards. Therefore, turning daily to our sources of guidance, in whatever scripture we accept, helps to maintain a constant learning process. Even if we have read the texts before, we always bring new experiences to our reading and take away new insights. Quality is more important than quantity. Reading one sentence with an open heart and mind can be more beneficial than a superficial or exhausting scanning of long texts.

Nature

The country is the world of the soul, the city is the world of bodies.[18]

Another path to spirituality is through contact with nature. The perfections of the Divine are reflected everywhere in nature. The beauty of a rose, the grandeur of a mighty tree, the majesty of a mountain, the splendour of a sunset can resonate with our inmost being and give us an intimation of what can be even greater. We only have to look for these reflections. The beauties of nature can also draw us out of ourselves and our sense of humility can strengthen before such wonders. We can come to appreciate that the source of all wealth is the earth on which we all walk.

Service

Like a tree whose purpose is to bear fruits, spirituality will not achieve its ultimate purpose if it is not put into action in our own life and a life of selfless service is an important way to acquire spiritual qualities. In the past, people seeking a spiritual way retired to a monastery or convent, or became hermits, devoting hours to prayer. But an active life of social contact provides many more opportunities to practice patience, forgiveness, detachment, humility, trustworthiness and other qualities of the spirit. A life spent seeking opportunities to be of service to others and to society can be very rich and rewarding and can make each of us a role model for the generations that come after us.

Building our spiritual reality is a life-long process, so what you learn while crossing this valley will always be with you. Spirituality is a state of mind and heart, not a destination, so we should not be over confident; we can slip and fall along this path at any time and lose everything. The ego, symbolized in ancient scriptures as Satan, the devil or the evil whisperer, is always looking for ways to exert itself. There is no point at which we can say we are 'saved' or have a guaranteed ticket to heaven. Selfish desire has reduced to ashes uncounted lifetime harvests of the learned.[19] In many ways, the higher we traverse on the mountain of spiritual growth and aim for the pinnacle, the easier it is to fall, as pride and self-satisfaction are spiritual pitfalls. Here, humility is an all-important antidote.

With the qualities you have acquired in crossing this valley of the self and climbing the surrounding mountains of selflessness while cultivating your spiritual nature, you are now ready to enter the valley of multiple higher realities.

6
The Valley of Multiple Realities

The valley of multiple realities, the sixth valley that you need to cross, is filled with an almost impenetrable fog. The dense mist blocks your view, so vision is of little use. Sound carries farther, but the mysterious sounds you hear are hard to interpret. You are disoriented as you try to progress, following what may be traces of a path and feeling your way forward. The only certainty is the earth beneath your feet, the physical reality that you are standing on. And this is the challenge of this valley: trying to understand reality; your own and that of the world around you and beyond you.

The problems of the world seem very real, as your experience of the lower valleys has shown. You can understand their origins and see their effects. They are important causes of human suffering, of life and death. You also have your own experience of reality. You know when you were born. You are alive. Your body seems very real to you. Other people around you also appear real. You can talk and play, love and fight with them. But their reality is less certain than your own. Do you really know what they think and feel? What if they were just the creation of your imagination, like in a dream, a vision or delusion, or perhaps even a ghost or spirit? What about virtual realities, as in films or video games? These are creations of others and their existence is recreated in your imagination and may be preserved in your memory. So, what *is* reality?

This is why this valley is known for its dense mist. We cannot see far without a guide. It is a wide valley, with many meandering streams and trails that cross back and forth; for the question of human reality is one that has troubled the greatest philosophers, mystics and scientists of all ages and epochs. This is one of those questions for which there is no definitive answer, but one that we must answer for ourselves in a way that satisfies us and that gives an orientation to our lives. We are called to examine questions such as:

Why am I here?
What is my purpose in life?
What will make me happy?
What happens when I die?

To find your way through this valley, you will need to draw on the two great knowledge systems that are science and religion, as each sheds light on this question from different angles. They are not contradictory, but complementary. In fact, as you will see, they are not so far from each other in many ways. This valley starts with the reality around you, before coming back later to the reality within you.

What does science say about reality?

At a superficial level, reality seems self-evident. Standing on the valley floor, the ground is quite solid. In fact, we take this reality so much for granted that if we experience an earthquake we may be traumatized because our faith in the solidity of the earth is tested. If you are sitting in a chair or on a bench reading this, the object beneath you seems quite real because it is holding you up. How is it possible to doubt material reality? But how can we prove that it is real? Is there objective proof that is the same for everyone? Is our own experience objective or subjective? We know that the mind can play tricks on us, as with mirages or optical illusions. How do we know that we are not just an avatar in some programmer's virtual world? Can our senses be trusted? Is anything real? Can we define reality? Can science lift the fog?

Suppose you are a scientist setting out to define reality.[1] You might start with yourself. You have a human body that eats, drinks, sleeps and is reading this right now. Your body was born, grew up, will age and finally die. You were not real before your conception, although you originated in another reality – your mother's womb – and not from nothingness and you will no longer exist as a physical body after your death and the dispersal of your molecules. Science gives you concrete tools to define your physical reality. However, you might have more difficulty proving that your consciousness is real and that your thoughts and feelings are emergent properties of your living body. And after your death, are the memories that others have of you real? What about their pictures of you, the recordings of your voice, or things that you made or wrote? Already at this level, what is real is not so clear.

Next, take the chemical perspective on reality. There are a limited number of chemical elements or atoms that combine in various ways to make molecules that make matter in all its various forms. For an atom, the number of protons and neutrons in the nucleus and the number and position of electrons spinning around the nucleus determine to which chemical element it belongs and how it combines into molecules. However, these subatomic particles are very tiny and most of each atom is empty space. Chemical reality is very different from what we experience physically.

What about reality in terms of physics – the dimension of reality that physicists study? Einstein showed that energy and matter are part of the same reality and can be converted from one into the other. Physicists have found all sorts of subatomic 'particles', including both matter and anti-matter that annihilate if they meet and they see evidence of what they call dark matter and dark energy making up 96 per cent of the universe without knowing what it is. Is this reality, or just theory because it has not yet been proven?

At the quantum level, reality is even more unreal. According to the uncertainty principle, quantum entities can be in superpositions of two states at once, or two places at once and are only probabilities until they are observed or measured by instruments that we make. Are they only real when they are observed? Is the reality in the quantum entity or is it in the instrument that observes it? Quantum entities are believed to pop into and out of nothingness all the time. They may be entangled, so that observing one determines the state of the other even though they are far away and no signal could possibly pass between them.

As instruments become more powerful, our discoveries of what is real become smaller and smaller and further away from what we normally experience as reality. There are scientists who believe there are more than the three dimensions of space and one of time – perhaps even ten – with the others rolled up. Some say that the ultimate reality is vibrating strings of energy that make up everything else. Others suggest that time might be able to go backwards as well as forwards. Cosmologists consider that we may be in only one of an infinite number of parallel universes that we shall never be able to observe. Scientists accept, explore and argue about these things as part of reality because they can be expressed mathematically or account for certain observations and hopefully experiments can be designed to explore their properties.

Mathematics is, in fact, the next level that science considers as real. It is a language that can express much that we know about the world. Mathematical formulae work. Numbers can be coordinates that tell where something is in space and in time. They can be organized into arrays of sets. At the same time, mathematics is a pure abstraction. Do numbers exist as a physical reality? Mathematicians devote their lives to exploring many forms of mathematics, proving that certain relationships can or cannot exist and admiring the beauty of mathematical expressions. Our society would be much poorer if we denied the reality of mathematics, even if it is beyond all tangible form. At the same time, is reality confined only to those things that can be proven mathematically?

Science can easily fall into the trap of reductionist thinking, breaking everything down into smaller and smaller entities, but losing sight of the properties of the whole. Systems science explores the emergent properties when everything is put back together and which often cannot be predicted from knowledge of the parts. Plato already considered things that exist independently of matter or mental entities, or concepts and ideas that exist only in minds. Either what is fundamental is not material, or nothing at all is fundamental. What then should be considered a scientific proof that something is real? Do we limit ourselves to only those things that can be observed and measured with instruments or described with mathematics?

One recent attempt to explore scientific definitions of reality identified the following[2]:

1. Reality is everything that exists without us, untouched by human desires and intentions. In this case, languages, wars and financial crises do not exist.

2. Reality is those fundamental things that everything else depends on: molecules, made of atoms, which are made of electrons and a nucleus. The nucleus is made of protons

and neutrons, which are made of quantum entities (and the other 96 per cent that we do not know, such as dark matter, dark energy, gravity), which are made of some still unknown foundation. In this case, higher-level entities, like Mount Everest, do not exist. As Heisenberg the famous physicist said: 'The ontology of materialism rested upon the illusion that the kind of existence, the direct actuality of the world around us, can be extrapolated into the atomic range. This extrapolation, however, is impossible . . . Atoms are not things.'[3]

3. Everything is reduced to mathematical reality. The universe is made of mathematics. But what is mathematics made of? The empty set 0=0; so nesting nothingness produces all mathematics. It does not require a physical origin or form, or space or time. Quantum particles are wave functions. As Galileo said: 'The book of nature is written in the language of mathematics.'[4]

4. Beyond mathematics, it is possible to consider information processing as the root of everything. Quantum information can spontaneously come into being. Interactions between particles produce binary answers, so the universe can be considered a quantum computer. The universe is ultimately information, held at boundaries or projected like a hologram.[5]

5. Does consciousness create reality? If nothing is real until it is observed or measured, the conscious mind selects quantum possibilities, making them real. It has been suggested that the emergence of a conscious mind created our universe from the multiverses.[6]

6. If we know that something is real, what do we know? And what do we mean by 'know'? Our senses can deceive us. Plato said that knowledge is justified true belief. For Descartes, his only certainty was that there is something that is doubting everything; thus one's own consciousness is all that there is. For the dualists, mind and matter are distinct; while in panpsychism, all matter has mental properties. Are definitions of reality just circular reasoning of statements about perceptions and statements about statements?[7]

7. With modern technology we can create all sorts of artificial worlds. If we can simulate reality, are we the basement level of reality, or are we in someone else's simulation? How do we know?[8]

We can, in fact, follow the above inventory of realities to its logical conclusion. With increasing abstractions from quantum physics to mathematics to information and consciousness, the logical next step would be the ultimate reality of perfect knowledge and perfect consciousness, encompassing all the other levels as less-than-perfect projections of that ultimate reality. Conversely, everything that exists at lower levels must find its perfect expression at this ultimate level. Since we are ourselves imperfect, we have to admit that we can never know or understand such perfection, which would be, by definition, unknowable by anything below its own level. This would bring an admirable dose of humility to the scientific enterprise, which too often assumes that it will ultimately explain everything and discover all truth. Human knowledge, including science, is inherently imperfect. We can continue indefinitely to explore the unknown and find rational explanations for everything in the universe, getting closer to that perfection without ever reaching it.

This is a scientific truth. No sequence of finite objects can more than roughly approximate an infinite one. Anything short of infinity itself comes nowhere near describing infinity. It follows, then, that ultimate perfection (infinity) is by definition beyond the reach of our finite understanding. In the debate between those who think nature is fundamentally mysterious and those who think that it is fundamentally intelligible,[9] both are right. The physical world described by science is fundamentally intelligible, but science itself leads us to acknowledge that our understanding cannot extend beyond a certain limit to encompass absolute perfection. We can acknowledge rationally that there must be absolute perfection without being able to understand it. If this valley is to bring you wisdom, then this is the first step.

Science and religion

The second challenge in this valley is to find the harmony between science and religion. In Western civilization, science and religion have often been in conflict. Galileo's scientific observations of the solar system conflicted with church dogma. Intellectuals like Voltaire were highly critical of religion, as expressed in a church that was out of touch with its times. Nineteenth century France saw the separation of church and state, and the establishment of a secular public school system. Today there are scientists who call for the abolition of religion as a medieval superstition that is dangerous to society, while on the other hand fundamentalist religious movements attack scientific efforts like vaccinating children against polio, or the teaching of evolution in schools.

But history also shows times when religion and science worked together. Islamic civilization preserved scientific knowledge from antiquity, created Arabic numerals and algebra, and made significant advances in astronomy and medicine. It was from the great Islamic civilization that scientific knowledge returned to Europe during the Renaissance. There is no inherent contradiction between science and religion.

A major part of the fault lies with religions which, though inspired in origin, accumulate over time the same faults as all other human institutions: irrational beliefs, narrow interpretations and dogmas, a literal understanding of metaphorical texts and clergies and institutions which build and then cling to power. The social teachings of a religion are also adapted to the time and place of their original foundation and may no longer correspond to the needs of an evolving society.

Religions were never intended to be permanent; they have all prophesied a time of return and renewal. When the form becomes more important that the substance and spirit, it is time for religious renewal. In fact, from an evolutionary perspective, all the major religions can be seen as one religion, part of a common process of the spiritual education of humanity.

With the values they bring and the laws they establish, religions could be compared to the basic operating system of a computer, determining the rules by which all the operations and calculations take place and how they interact. As computer systems evolve and become more complex, it becomes necessary to update the operating system, or else it will hold back further improvements.

Religion, in this larger sense, freed of its sectarian trappings, is a necessary complement to science. Science, for all its accomplishments, takes no moral or ethical position. A scientific discovery can be used for good (nuclear medicine) or evil (nuclear weapons). Science serves totalitarian states as well as democracies. Science needs a moral and ethical complement to avoid falling into a narrow materialism and religion and spirituality can provide that. Religion, without the support of the rational tools of science, too easily falls into superstition. Both are approaches to the same ultimate truth and thus have no inherent contradiction. Science helps us to understand and benefit from our material existence, while religion gives us a purpose in life and shows us how to use science for our benefit. Together they can build an ever-advancing civilization. You should feel free to rely on both on your journey through life.

The limits of science

One of the barriers that you will come upon in the mists of this valley is the wall limiting the reach of science and the questioning of what science can establish as real. One scientist, considering another inflationary cosmologist's mathematical description of infinite parallel universes as the ultimate demonstration of reality, asked if this was still science or veering towards something akin to religion, with uninhibited scientific speculation requiring the same leap of faith as believing in a Creator. Science involves both speculation about what is rational or possible and skepticism about what is scientifically proven or established as the best explanation available.[10] This raises the question about the boundary between science and other domains of human experience. What do we do with questions that cannot be proven with material data or experimentation? Do we deny the reality of everything that is beyond the reach of science?

Rather than dogmatically limiting reality to only those areas where scientific experimentation can produce material proofs, the tools of rational thought at the heart of science can also illuminate other less tangible domains of human reality. We need, however, to identify other standards of proof, perhaps more abstract or more personal than those of science, but nevertheless within the scope of human reality and rational thought. A purely materialist perspective excludes too much that is real and important in human experience. As is apparent in the efforts of science to define reality described above, there is almost always a leap of faith or belief that accompanies the rational intellectual process, or that underlies the circular reasoning – as was acknowledged by Plato.

There are two extremes we need to avoid: on the one hand, a definition of scientific knowledge that is limited to material reality and denies anything beyond the human being as a biological entity, with only material needs to be satisfied to be happy; and on the other hand, irrational speculations or imaginary constructs of 'reality' that fall in the domain of superstition. Both can be equally destructive to the fulfilment of our human potential, leaving us too limited in what we try to do with our lives.

What then, are some of the domains of human reality that escape from material proofs? Obviously there is what could be called our **intellectual reality**, the world of ideas and thoughts, of theories and philosophies, of world views and representations of what we experience. There is also our emotional reality – how we feel about things that resonate with us at some deep level – that can be defined as beauty, love, empathy and compassion.

Those things we call art, whether literature and poetry, painting and sculpture, music, theatre, dance and other components of culture, usually combine both intellectual and emotional dimensions and communicate with us at multiple levels. These are clearly real, if not universal. Different people react to them in different ways, depending on experience, upbringing and personal preference. These things are not beyond rational proof. We can explore how they communicate to us and why we respond the way we do to them. And people can be classified by their different responses, for example as lovers of classical music as opposed to heavy metal.

Some things elevate us and bring out our best qualities, while others play on our animal emotions and can inspire violence or aggression. However the scientific method does not tell everything about these things. A mathematical proof can be beautiful even if it says nothing about material reality. The Mona Lisa's expression speaks to each viewer at some deep level beyond the paint on the canvas.

It is the third level of human reality, **spiritual reality**, that is the most difficult to define and that challenges those with a more materialistic perspective. On the one hand, most people in the world belong to some kind of religious or spiritual tradition and would never question the existence of their spiritual reality. On the other hand, many scientists for whom material proof is an absolute requirement and others who have rejected a religious tradition as irrational superstition (often with good reason), will deny the existence of anything 'spiritual'. This issue can raise strong emotions on both sides. We thus need to look rationally at what is behind spirituality.

Religion

As explored at the beginning of this valley, science and religion can be considered as two complementary knowledge systems. But first we need a clear vision of what religion really is or should be. While the advantages of scientific knowledge are well understood, there are more doubts about religion – which today is mostly associated with division, strife and repression, if not terrorism, rather than a source of knowledge.

Much of what is presented as religious knowledge is not in harmony with science and much of what is presented in the name of science denies the spiritual capacities cultivated by religion. It is important to remain free of simplistic and distorted conceptions of science and religion and of an imaginary duality between reason and faith. Reason should not be confined to the realm of empirical evidence and logical argumentation and faith only associated with superstition and irrational thought. Our understanding of spirituality should be reasonable and science must recognize the need for a moral framework for its proper application.

Any process has to be rational and systematic, using scientific capabilities of observing, measuring, rigorously testing ideas and, at the same time, be deeply aware of faith and spiritual convictions and their role in motivation. Faith and reason can best be understood as human attributes through which insights and knowledge can be gained about the spiritual and the physical dimensions of existence. They make it possible to recognize the powers and capacities latent in individuals and in humanity as a whole and enable people to work for the realization of these potentialities.[11]

Religion has for many become a negative word, associated with the rise of fundamentalism, intolerance, superstition, terrorism and violence. Unfortunately, today there are movements that call themselves religious with all these characteristics, often created in order to pursue political or ideological ends. Science has also been misused to pursue political ends, as in Nazi Germany.

Setting aside all the labels and asking what is universal about religion as a phenomenon of human society, there are a number of defining characteristics: the acceptance of a spiritual dimension to human reality, which can be cultivated through education and individual effort and which, being immaterial, persists after death; a set of moral precepts about what is right or wrong ('Thou shalt not kill') and ethical principles to guide social interactions ('Do unto others as you would have others do unto you'); and belief in a higher reality, Absolute Truth or Unknowable Essence, variously called God, Allah, Jehovah, etc. to whom we owe allegiance and must learn to love. These will emerge again as you move up this valley.

Is spirituality another dimension of human reality? And is

religion an educational process for cultivating it? It cannot be proven the way carbon atoms can be proven, but it produces effects in human behaviour and the building of civilizations, just as dark matter is assumed to exist because of its gravitational effects on the behaviour of galaxies. Is there a rational basis to establish the reality of human spirituality? For the religious believer, its existence is a matter of faith; just as for an atheist, its non-existence is often equally a matter of faith or belief. One can take an historical or anthropological perspective and say that the persistence of religion and other forms of spirituality across many epochs and cultures proves that it represents a significant part of human experience. Even today, those that deny its reality are in the minority.

A second way is to look for people who have followed a religious or spiritual path and see if they have qualities that others do not have. Is their spirituality expressed in the values they live by, the way they relate to others and their personal example? Of course, many people say they are believers or claim some spiritual station but do not practice what they preach, or even use their declared position to acquire a following, collect money or seek political power. It is not always easy to distinguish a genuinely spiritual person from an impostor. True spirituality is reflected in humility and self-effacement, so it does not broadcast its presence. We have to search it out.

Perhaps the most direct approach is to use our own personal experience. While we may be born with a potential capacity for spirituality, just as we may have a capacity for athletic performance, it must be cultivated. We can learn about it through sacred texts or scriptures and develop it through exercises like meditation, prayer and fasting. We may find role models in people who are further along in their spiritual development than we are. Obviously, if someone has always denied the existence of spirituality and has never tried to discover such a potential in themselves, it is easy to say that it is a superstition without any basis in reality. The objective way for us to test the reality of spirituality is to try it with an open mind and heart.

One feature that is common to almost all spiritual and religious traditions is the struggle between our lower material or animal nature and our higher spiritual nature – symbolized by hell and heaven, the devil and the angel, or egoism and altruism. Religions provide role models and examples of people with exemplary spiritual development as saints, gurus, elders, or those more recently admired for their self-sacrifice in service to others, such as Nelson Mandela, Mother Teresa or Gandhi. Religions teach that our true reality is spiritual and our purpose in life is to cultivate that reality. Our spiritual reality, or soul, while taking form in this existence, is not limited by it and continues after death in ways we cannot imagine – because we have no experience of an existence without time or space.

As you saw in the valley of individual discovery, the struggle with the ego is never over. It is part of our lower or animal nature that is always lying in wait, looking for some new way to express itself. In many cultures and religions this ego self takes on the traditional symbolism of the devil or evil whisperer, the malign spirit that takes possession of a soul. It is reflected in many things that were classified as sins long before modern society gave them commercial value: pride, jealousy, envy, greed, selfishness, narcissism, hedonism, seeking power and wealth and corruption and lying, among others. Notice that these are things that divide people, that prevent healthy social relationships, that undermine trust and confidence and that prevent the building of a solid community and society. They are counterproductive in achieving a collective human purpose.

Religion provides the tools to master the ego. It emphasizes love as a positive force of attraction: love for other human beings, love for your enemy, love for God and universal love.

It shows the potential of love to counter the ego and to replace hate with a stronger feeling of love. It can help in all the stages in the development of love: from love of self to love of family, friends, spouse, children, culture, nation, all of humanity and ultimately love of the Unknowable Essence that we call God.

The ego cannot be mastered without an alternative focus of attention and effort outside of ourselves. Self-love needs to be turned in some other direction. This cannot be other people because we are all full of faults and not always very lovable. For our growth to be unlimited, we need an unlimited point of focus or direction. This leads us to another essential concept and challenge on this path, which lies at the heart of most religions and is generally referred to as 'God'.

God

The concept of God provides this essential external focus for our evolution, individually and collectively. Once we accept that there are other realities beyond material reality, we can naturally ask how far these realities go. This opens the potential for endless realities and then, given our limited minds, it is natural to ask if there is an ultimate reality. We read above how science has taken the same path and has stopped just short of an absolute, ultimate reality that encompasses all others.

Religion is the domain of human experience wherein to ask such questions and it provides a source for this concept that has evolved as human understanding has advanced. Depending on the specific religion and culture, this concept has been given multiple names: God, Allah, Jehovah, Nirvana, Dieu, Gott, Dios, Deus, etc., each with baggage of various depictions and interpretations. It is important to recognize that the concept is inherently limited by our own understanding and by the available languages used at different stages in our own cultural evolution. Since these interpretations always fall short, it is easy to reject them and to claim that 'God is dead', or be agnostic (refusing to take a position on the issue of God) or be atheist (denying the existence of God).

The rational solution is to accept that the term 'God', or its equivalents, applies to Absolute Perfection, the Ultimate Reality, the Unknowable Essence – which is beyond names and attributes, coming and going, time and place and physical existence and non-existence. No words that we use can describe such an Absolute and no idea or conception of our mind can encompass it. Like infinity, which is beyond numerical representation, God is beyond any kind of representation, while symbolizing the perfection in all of them.

Since no effect in the reality that we know can exist without a cause, God could be considered the Ultimate Cause. Every characteristic that we can observe in the universe or ourselves, including abstractions such as love, knowledge, forgiveness and questions, must find its absolute expression in the attributes of God. We are forced by our language to personify 'Him', despite our knowledge that God is beyond any gender, and must acknowledge that it is our language that falls short, not God.

While we are inevitably frustrated by our inability to conceive of or define God, we can nevertheless try to love and worship this Absolute, as this draws us out of ourselves towards limitless possibilities. This is perfectly scientific and rational, since science also acknowledges a perfection that is unattainable, as well as potential realities that are beyond scientific understanding (such as before the big bang, or beyond quantum reality). Science can imagine or theorize about a material perfection while acknowledging that it can never exist in practice and thus is unmeasurable. The idea of spiritual perfection is a rational extension of this. The concept of God in this sense is not unscientific, although it can make materialists uncomfortable.

It is part of human nature to want to imagine or to create an image of anything we think about, including God. It is hard to admit that something is unknowable. It is even harder to love something that is unknowable and forever beyond reach.

Ye shall be hindered from loving Me and souls shall be perturbed as they make mention of Me. For minds cannot grasp me nor hearts contain Me.[12]

Yet, love, the ultimate force of attraction, is at the heart of spiritual reality. Our existence as created beings is a result of God's love for us, which we must return.

Love me that I may love thee. If thou lovest Me not, My love can in no wise reach thee.[13]

Love becomes pure when it is detached from self and all limited conceptions and it finds its ultimate expression in the love of God.

Our understanding of God and our conception of a path leading us towards God, will be constantly evolving and expanding throughout our lives. These concepts are highly personal, as each one must investigate independently within the limitations of our experiences and the expressions available in our language. In religious texts, the approach is often to use symbolic and metaphorical language for our relationship with God, such as 'the lover and Loved one', or being 'near to or far from God'. There are metaphors that describe God as being 'closer than your life vein' and referring to 'the ocean of [His] Presence'. The process of spiritual development, of drawing closer to God, is sometimes described as 'walking the mystical path' or 'passing through Seven Valleys'.[14] An unknowable God is beyond this journey, but our spiritual purpose is fulfilled through it. We must have the humility to accept that we cannot know or understand everything, that we can evolve and seek greater perfection only in our own human condition and that awareness of and love for that Ultimate Perfection can motivate us to improve.

The challenge we face is personification, the natural tendency to assume that God is like us and to use a personal – and culturally-based – vocabulary to describe Him. The negative effect is our tendency to bring God down to our own level, which is contrary to any concept of Absolute Perfection and ultimately results in the logical rejection of that personification – which is not the same as rejecting God. Personification can be positive symbolically, in that it helps us to relate to the need to grow towards that Perfection, to fulfil our own highest purpose through a better knowledge of our own selves and our potential to develop. When we ascribe attributes to God, like All-knowing, Forgiving, Generous, etc., they do not really describe God, but remind us of our own need to acquire knowledge and to be forgiving and generous. It is only important to remember this limitation; that everything we think, say and do falls short of this Absolute Perfection.

One collateral effect of the personification of an all-knowing God, Who is 'conscious' of everything we do, is His role as the perfect enforcer in keeping us honest. Research shows that people who know they are being watched are more honest. Knowing that God is always watching us and will ultimately confront us with our wrong-doings is a strong incentive to do what is right. Believing that the punishment of wrong-doings is not limited to this world thus contributes to building healthy social relationships and communities and to the progress of societies. While this function can be replaced in part by strong institutions of law and justice, they are less effective for those 'hidden crimes' that go undiscovered. This is in addition to the other roles of religion, of inspiring ethical behaviour and a constant effort towards self-improvement. It reinforces the fear of punishment and inspires good behaviour that would be pleasing to God.

Returning to the issue of human purpose, if our purpose includes acquiring good qualities and virtues and refining our character, then learning to know and to worship God means turning towards the ideals of human perfections – which are attributes of God – with love and appreciation. Love is a two-way process of giving and receiving. Learning to love the unknown and unknowable Absolute Perfection helps to turn positively to the unknown in ourselves, towards our hidden potential and to the unknown in others, helping us to overcoming prejudice and to the unknown in the world – which is the impulse for scientific research and exploration.

A related issue is the immortality of our individual spiritual consciousness, often referred to as the soul. Is consciousness just an emergent property of neuronal processes in the brain, or can it conceivably have an existence independent of the physical body and thus with the potential to continue to exist after death? Materialist science still hopes to prove the former. The great majority of religions and forms of spirituality accept the latter.

While our body is the instrument for our soul to acquire divine attributes that achieve our human purpose in this life, those attributes are intangible and nothing requires that they disappear on the death of the body. If there are realities beyond this physical reality and endless spiritual worlds beyond this world, then our emergent properties – such as consciousness cultivated with divine attributes – have the potential to continue evolving towards that Absolute Perfection that is God. Acceptance of a concept of God and acknowledging forms of existence, or worlds of God, beyond this material plane (in what might be called a Divine cosmology) are simply a higher form of reality.

Since science already hypothesizes multiple universes and multiple dimensions beyond those we experience, it is equally logical, or at least rational, to consider that higher levels of consciousness can lead to higher planes of existence beyond the physical dimensions of space and time. While these are beyond the present reach of the great knowledge system that is science, the best evidence comes from that other great knowledge system – religion – with the founders of those religions providing the evidence in their writings and their example of the existence of these higher levels of perfection, including God. We can understand the purpose and utility of belief in God and in immortality, while acknowledging the limitations of rational scientific approaches.

The soul can be considered as that dimension of human reality with the capacity to become conscious of the Ultimate Reality that is God, to accept the truth of spiritual existence and to turn to it in love, in turn reflecting divine qualities. Just as it has the potential to mirror these perfections, so can it equally become a victim of self and passion and sink to the lowest levels of human existence. There can be many stages in the development of the soul on its journey of spiritual growth towards higher realities, starting with its recognition of the latest Divine Educator that manifests the Absolute Reality and the acceptance of the spiritual principles and laws relevant to our own time. The soul grows through its efforts to put these into practice.[15]

The emergent processes in the brain that produce consciousness and feed the soul are intimately linked to our sense of place and movement, our perceptions of seeing and feeling, our will and purpose and our thoughts and speech, all of which depend on their association with our body and that end immediately if that relationship is cut. Those processes are the link to acquiring the higher perfections that are attributes of God and characterize the soul, which means that we can never fully understand the operation and potential of the soul. It is, in a sense, the presence of a reflection of that ultimate reality within us. The more we think about it, the more we must acknowledge our helplessness to understand it and that acknowledgement in humility is in fact the ultimate aim of our spiritual development.[16]

Manifestations of God

The gulf that separates us from the Unknowable Essence and Absolute Reality that is God is so great that it may seem impossible to fathom. Fortunately, religion also provides an intermediary between God and humanity in the form of individuals who can best be called Manifestations of God, whose function could be described as Divine Educators. The founders of the great religions (Krishna, Zoroaster, Moses, Buddha, Jesus Christ, Muhammad, the Báb and Bahá'u'lláh) are the ideal form of personification, in that they mirror the Absolute in our own human reality and provide a model to follow; just as a mirror can reflect the attributes of the sun, its light and heat, as we never can experience the sun directly.

Science advances through brilliant men and women who have a clearer understanding of physical reality, who design and undertake experiments to explore it and describe it for the rest of us in scientific principles and laws. Religion advances in the same way, with extraordinary Educators who reveal spiritual principles and laws. Take, for example, the law of attraction In physics, it is gravity or magnetism. In human terms, we talk about love: love between people, love of nature and, at the highest spiritual level, love for God and God's love for us. The laws of attraction are complementary, not contradictory.

Periodically throughout history, these unique individuals have appeared with an understanding of spiritual realities beyond physical reality – often termed 'worlds of God' – through a process which they themselves term a 'Revelation'. This is not the result of their own intellect, but a received inspiration of what humanity needs at a particular point in its evolution relative to that Ultimate Perfection.

These individuals manifest God as they translate that Absolute Perfection into the relative perfections of our own plane of existence. Since our human limitations prevent us from knowing the ultimate reality of God directly, these Educators manifest the qualities of God to make them accessible to us. They are the intermediaries between God and humanity, the source of perfect knowledge for our time and place in the evolution of human reality, providing the path for us to free ourselves from ego and to take a further collective step in our human evolution. They also provide principles and laws that we must obey to fulfil our potential and to take civilization forward and they set an example in their own lives of how these principles should be applied.

The religion they each establish provides instruments and institutions to help society advance at a particular stage in its evolution. As society progresses, these laws and institutions need to be adapted to the changing requirements of each succeeding age. All religions are part of the same process of human social and spiritual evolution, periodically renewed through progressive revelations as our needs change. Religion evolves and goes through cycles of what systems science calls a punctuated equilibrium, with a creative impulse producing rapid innovation followed by consolidation towards equilibrium and then decline, just like the rest of nature. As history shows again and again, this process of religious renewal, as with any major change, tends to be rejected and denied at first. It takes humility and detachment to accept the Manifestations of God in each new new form and each with a new name. Our egos, attachment to the past, to power and status and to prior knowledge, are barriers to acceptance.

What, then, is the rational evidence for the reality of this process of religious revelation? Materialist intellectuals find it easy to discount the evidence behind religious traditions, as dimly reflected in history and buried in generations of interpretation and mythicization. What can we really know about the life of Jesus, for example, if the stories of that life were only written down generations later? As in archaeology, where we try to understand the reality of an ancient civilization from the ruins of some settlements, the trash in ancient rubbish pits and the occasional burial, it is easy to dismiss the evidence of something as intangible as an ancient Revelation or set of spiritual teachings.

This is one reason why religion has to be renewed from age to age. While the persistence of old religions is one evidence of their residual power, the real proof of the process of religious renewal should be sought in the most recent and best documented example of this in action.

A recent example that is well documented, and that we can rationally study and investigate independently, is the Bahá'í Faith and its founders, the Báb (1819–1850) and Bahá'u'lláh (1817–1892). This is not the place to go into detail about this example, but many sources are available for those who are interested. It is sufficient to say that their lives and teachings are well documented and they renew religion as a positive force to address the needs of the modern world. Understanding the positive role of religion is another important step forward in your journey through this valley.

Wellbeing and happiness

With this new perspective on the complementarity of science and religion, what are its implications for a broader definition of our human reality and purpose? Why are we here and what should be our goal in life? For our physical wellbeing, we need to live in a healthy, sustainable and preferably beautiful environment and science is our best guide for this, even if contact with nature has a spiritual dimension that can draw people out of theirselves and uplift them. For our social dimension, both science and religion have essential things to say about justice, equity and social organization. We all surely want to live in a peaceful society where everyone has a role and place in supportive communities that are united in all their diversity.

To redesign our economic system, we need a society that starts with spiritual principles of justice, cooperation and altruism, giving everyone the opportunity to work and eliminating extremes of poverty and wealth as the design principles for new economic mechanisms and structures. The world has now been unified for the first time by science and technology, especially in transport and communications, leading to a global economy. We are now challenged to accept the oneness of humanity, living with everyone in a global community. Religion shows us the path to live in peace and unity as a global family and calls on us to consider the wellbeing and true happiness of every human being as a trust of the whole.

In complex evolutionary systems, achieving each level of integration opens the possibility of new higher levels. A certain level of functioning at lower levels is necessary, but not sufficient, to fulfil higher levels. The measures of success are different at each level. There are certain optimum values for performance and efficiency at each level and going beyond these to an extreme can be damaging. Human wellbeing is the same. We need a certain level of physical wellbeing to have the means to invest fully in social, cultural and spiritual wellbeing. However, an excess of wealth and material wellbeing can reinforce the ego and make spiritual wellbeing more difficult. Life includes unavoidable problems like illness, bereavement and ageing, which help us to become detached from material and social wellbeing in order to greater strengthen our spiritual qualities.

True happiness comes from fulfilling our higher purpose by refining our character and acquiring spiritual qualities and by contributing to an advancing civilization. Happiness is not a goal to be worked for but the result of our efforts and struggles to achieve other goals, in effect looking back on a life well lived. In the afterlife, we shall presumably be even more conscious of what we have accomplished in this life. Through this, our complex evolutionary progression continues towards higher levels of perfection – towards God – but in ways that are beyond our ability to comprehend.

Religion in its pure form, then, is the organized system that channels to all of us the knowledge of that Ultimate Reality that is God and that gives us the laws that we must follow to achieve our higher purpose. It renews those universal principles at the heart of all religions and provides the social laws necessary for our further collective evolution. Religious truth is relative to a particular time and place and to humanity's stage of development, which is why it becomes harder to find fulfilment and happiness in old religions that have outlived their usefulness. The world of today is as far from the worlds of Moses, Buddha, Jesus or Muhammad as the airplane is from the chariot.

Cultivating spirituality in society

Religion not only provides a definition of higher human purpose and a model to follow, it also furnishes tools to help us to cultivate our spiritual qualities. Some of these tools you will have encountered in the previous valley. Religion also provides collective tools to support our spiritual advancement in unity with others.

One feature that religion brings to social organization is a form of worship, or collective turning towards that Absolute

Reality that we call God. Making devotion a social act adds a spiritual dimension to social relationships. When we share something so deep within us, it strengthens our bonds with others and reinforces our sense of community. The specific forms that worship may take vary with the religious tradition and with time may become formalized in rites and rituals that can lead to disagreement and separation – despite their essential purpose to reinforce unity. Simple worship in community devotional gatherings can easily create ties across different forms of spirituality. Returning to this simple form of shared spirituality can be an important part of community transformation.

In the previous valley you learned about the challenge of mastering the animal side of our nature in order to grow spiritually. One powerful practice for learning the benefits of material sacrifice is through financial giving, Zakat or tithes, where material wealth is turned to a spiritual or social purpose. If we truly believe that our spiritual side of life is really more important than our material existence, than it is only consistent that we share our wealth in support of that dimension of life and society – preferring others to ourselves.

It is part of human nature to want to express our feelings and words are often poor vehicles for this. So we often turn to music, song, dance and art for a more complete expression of our deepest emotions. The result has been a rich flowering of culture, through music, art and architecture. Music, in its most refined form, can be the language of the soul, lifting our spirit and giving wings to our prayers. Our houses of worship may also aim to be places of beauty, visually uplifting and inspiring. Creating a work of art that touches the souls of others is a wonderful service to humanity.

In fact, service itself is an important path to spirituality. This is not the time to withdraw to a monastery and cultivate spirituality in isolation. True growth comes from serving others. Even work, when performed in a spirit of service, aiming to do the very best of which we are capable, is a form of worship. A life of service to others is a path to spirituality. A community in which everyone serves others will advance to higher and higher levels. The highest aim of religion is to empower an ever-advancing civilization. What better way is there to cross this valley and to climb up to the heights of individual and collective spiritual fulfilment?

7

The Valley of Visions of the Future and the Mountains of Innumerable Possibilities

As you enter the seventh valley, the valley of visions of the future, you are approaching a high point from where you can leave the chaos of the present behind and stretch out towards the future. You should now be equipped with a clearer understanding of where you have come from and why the world is in the mess it is in today. It is hard to be motivated when you are lost, wandering in a wilderness with no idea about how to get out of it. Something needs to give you a sense of direction, to keep you from just going in circles. Ideally you would want a map and a compass, or a GPS to show you where to go and to keep you headed in the right direction. And you also need a destination, the place you want to reach. It is the same with life. People who have a clear vision of the world they want to build or the society they would like to live in are more effective in using their talents constructively for a positive end.

You probably now know yourself a little better than before, even though discovering your true self is a lifelong process. You will hopefully have a clearer sense of your human purpose, why you are here and what things you can do to refine your character and to achieve the real satisfaction that comes from being of service to others.

The challenge of this valley is to decide what you will now do with all these things that you have acquired. And for that you need to know yourself, your talents and interests – or at least be open to discovering them, even if this means taking risks and accepting mistakes as a normal part of the learning process. Then you need to construct your vision of where you want to go. This valley is surrounded by many peaks, representing innumerable possibilities about the future. Reaching a summit will be the journey of a lifetime; but which summit? How can you choose? And then, how do you find your way?

Of course, the future is unpredictable and there will be many surprises in your path – from events in the world and from your own encounters and discoveries. Destinations frequently evolve or change entirely as new possibilities emerge. Christopher Columbus set out for one world and discovered another and you may well find the same happens to you. This does not really matter if your direction of travel is set by your values. They will bring a coherence to your life. If your goal, for example, is to use your talents to be of service to others or to society, many different paths can be equally rewarding and no single path is better than an other.

We saw, in the third valley, the role of social cohesion in the rise and fall of civilizations. The wealth of a growing population and the improvement of technology produces a successful elite, until excess population allows increased exploitation of labour and an overshoot of natural resource limits – wherein the poor suffer and the elite continue to live well. A generation later, the excessive concentration of wealth leads to conflict among a too numerous young elite over a shrinking resource base and the civilization loses cohesion and collapses.[1] The same process is happening today with educated youth who hoped for a better world that they do not see coming, or who no longer have access to the success and comforts of their parents' generation. With other ways blocked, they are being pushed towards revolution and risking factionalism and anarchy, all of which have been predicted to lead to political instability and impending crises in Western Europe and the USA. The only way to avoid this would be to reduce social inequality.[2]

Movements of frustrated youth around the world and their attraction to antiestablishment factions, suggest that such pressures are building and the possibility of a widespread youth rebellion cannot be excluded. The main problem is that while many youth know what they want to tear down, they lack a clear idea of what to put in its place. Revolution that leads to anarchy and dictatorship is counterproductive.

Many other people of all ages are similarly frustrated and angry. While the world economy has generated massive new wealth and technological marvels, not everyone has benefited. Many have lost their jobs and not found replacements, or have seen their income stagnate, or even shrink, while prices keep rising. The wealthy in urban areas may have benefited, but those living in slums or rural areas have little hope. Whole countries have suffered from poor governance, corruption, division and civil war and many millions have become refugees. Globalization has seen more losers than winners. Many are drawn by fear to populist and antiestablishment movements, or are attracted to autocratic leaders. Even those with the best motivations find it hard to maintain hope and to search for a positive way forward.

All of the valleys you have crossed in your journey have shown the forces of disintegration that are tearing down a dysfunctional world system and are helping you to discover the forces of integration that are laying the foundations for a new world civilization to emerge. Rather than wasting your time and risking your life to tear down the present system, why not start building the new one that must take its place?

This seventh valley has a dense forest of good intentions, in which it is only too easy to get lost. A life of good intentions without real action ultimately leads nowhere and is not very satisfying and may do little to fulfil you or to advance civilization. At the same time, there is no single path out of this valley; there are as many paths as there are human beings. Each of us is different, with a unique combination of qualities and potentials and also weaknesses to overcome. You have to set your own direction and make your own path. If you have a clear vision of your goal, or at least your direction of travel, your path will be consistent; if you are unsure, your way will certainly wander, will take much longer and not go so far.

This also means that there is not one single vision of the future that everyone can simply adopt. Yes, we may agree in general terms that a world of peace and justice is desirable, but every individual must find their own specific vision of their place in building that future. There is not one single mountain ahead that will take you up out of this valley, but a whole mountain range with many peaks and ridges, lakes and meadows and passes from one summit to the next, where you can search for the life path that suits you best. In this valley you can discover some of the options, which you can mix and match to find the vision of the future that fits you and satisfies your specific needs and longings. The lessons learned throughout your journey will be the best provisions for your life ahead and will give it the best chance of being purposeful and satisfying.

Taking charge of your own life

Do not think that you are powerless if you are young and inexperienced, or without material means or access to power. You are also the future. Every great tree starts as an insignificant seedling, easily bent in the slightest breeze. We know that to flourish, we need to reach towards the light. For a tree, it is sunlight that makes growth possible. For us, it is the spiritual light of guidance, the warm love that we receive if we turn our own love outwards. The world is full of dangers, distractions and traps to be avoided, like the storms that can break branches and even uproot the tree. Building a strong trunk of good qualities and sheltering branches able to carry good fruits, can protect us against the tempests of life.

Similarly, those at any stage of their lives – perhaps tired of its emptiness, struggling to overcome a disappointment, facing a dead end or a turning point, or even confronted by the inevitable decline of old age – can take hope that change is possible, even inevitable. It is never too late to join in the constructive forces building a better world.

The dangers to be avoided

There are three dangerous passes in this valley, where a slip and fall can send you tumbling back into the valleys below. The world is assailed by an array of destructive forces which can sweep you away in the deepening confusion, producing a sense of hopelessness, when instead we should be driving progress.

The first is materialism, rooted in the West, that has now spread to every corner of the planet. In the name of a strong global economy and human welfare, materialism markets a vain and foolish worldview founded in a culture of consumerism. It skilfully and ingeniously promotes a habit of consumption that seeks to satisfy the basest and most selfish desires, while encouraging the expenditure of wealth so as to prolong and exacerbate social conflict.[3] As you saw in the previous valleys, this consumer society is raping the earth, destroying nature and flaunting a material success for the few that is out of reach of the masses of the poor. Conflict is the inevitable result. We are living in an age consumed by self-interest, in which even spiritual matters are weighed upon the scales of reward and personal satisfaction. We are all the targets of this aggressive materialism that is intent on making us passive consumers, addicted to its fads and products.

The second danger is the rising tide of fundamentalism, bringing with it an exceedingly narrow understanding of religion and spirituality. It continues to gather strength, threatening to engulf humanity in rigid dogmatism. In its most extreme form, it conditions the resolution of the problems of the world upon the occurrence of events derived from illogical and superstitious notions. It professes to uphold virtue, yet, in practice, perpetuates oppression and greed. Fundamentalism leads to a myriad social problems, to intolerance and division and even to violence, conflict and terrorism.[4]

Third is the lack of any commonly agreed moral code or sense of right and wrong, as old standards have eroded and the hypocrisy of paying lip service to high ideals while acting in fundamentally contrary ways has become widespread. We are thus surrounded by the indecencies, the vices and the false standards which today's inherently deficient moral code tolerates, perpetuates and fosters everywhere. The external forces at work on our hearts and minds are pernicious indeed. Exhortations to live pure and chaste lives are not sufficient. Our minds are also affected by the choices our parents made for their own lives and often, no matter how unintentionally or how innocently, such choices condone the passions of the world. Its admiration for power, its adoration of status, its love of luxuries, its attachment to frivolous pursuits, its glorification of violence, and its obsession with self-gratification. We must realize that the isolation and despair from which we and our friends suffer are products of an environment ruled by an all-pervasive egoism and materialism. Our challenge is to replace this present-day order. This is the last mountain you have to climb, armed with all the qualities you have acquired on your journey thus far.[5]

The power of a positive vision

Those of you who are still in your youth should recognize that you will inevitably inherit the future. Today's young people will be tomorrow's leaders. Youth are typically in the vanguard of change. From their mid-teens through their twenties, most youth are not yet encumbered by a fixed career path or family responsibilities. You have an energy and mobility that gives you an exceptional potential for selfless service. You can lead the future progress of society and the expansion of new and better ways of living. You can easily recruit other youth to join you in meaningful acts of service and community building. You are eager to take on a measure of responsibility to aid the spiritual and social development of those around you, especially ones younger than yourselves. You can resist the temptation of self-interest, material reward and personal satisfaction and adopt a new positive vision of the future, putting the needs of others before your own. Through the example of your own efforts, you can contribute momentum to change within the whole community.[6]

Every generation has an opportunity to make a contribution to the fortunes of humanity, unique to their time of life. For you, now the moment has come to reflect, to commit and to steel yourself for a life of service. Rising above a distracted and bewildered humanity, you can adopt a clear vision of the possible future and demonstrate an integrity and uprightness that are not undermined by dwelling on the faults of others and without being immobilized by your own shortcomings. You can bring those who have been excluded into the circle of your intimate friends. Your consciousness of the failings of society can impel you to work for its transformation, not to distance yourself from it. Your motivation can be so strong that, whatever the cost, you will refuse to pass by inequity in its many forms and will work, instead, for justice for all of humanity.[7]

For this to happen, you need a fundamental shift in perspective, explored through all the valleys on this journey, that changes the way in which you view certain essential concepts: the true purpose of life, the nature of progress, the meaning of true happiness and wellbeing and the place that material pursuits should assume in your individual and family life.

With this new perspective, you should be convinced of the nobility of human beings; eagerly seek a deeper understanding of the true purpose of existence; able to distinguish between truth and superstition; clear in the view of science and religion as two independent yet complementary systems of knowledge that propel human progress; be conscious of and drawn to the beauty and power of unity in diversity; secure in the knowledge that real happiness is to be found in service to your community and nation and to the peoples of the world; and mindful that the acquisition of wealth is praiseworthy only insofar as it is attained through just means and expended for benevolent purposes, for the promotion of knowledge and toward the common good. The qualities you should cultivate include modesty, purity, temperance, decency and clean-mindedness. In this way, you can prepare yourself to shoulder the tremendous responsibilities that await you.[8] While all this seems ambitious, it is within your reach – if you are prepared to make the effort.

As you consider your place in the economy and look for work, you should distance yourself from the atmosphere of greed that surrounds you. You should consider the difference between gaining wealth through earnest effort, in fields such as agriculture, commerce, the arts and industry, on the one hand and, on the other, obtaining it without exertion or through dishonest means. What would be the consequences of each for your spiritual development, as well as for the progress of society? What possibilities exist for you to generate income and acquire wealth that will ensure your true happiness, both in this world and in the next? The goal should be the development of your spiritual qualities, such as honesty, trustworthiness, generosity, justice and consideration for others and the recognition that material means are to be expended for the betterment of the world.[9]

Including hope in your vision of the future

Having gotten your own life in order and your priorities set, you need to focus on positive visions of the future and to see beyond the mess we are in today. From a systems perspective, the potential of the future is so great that you cannot imagine what it might hold. However, you can identify the ethical principles that produce social cohesion and the processes of civilization-building that should be encouraged. Most importantly, you can concentrate on the efforts that you can now undertake to experiment with new forms of social organization, new processes of wealth creation and new approaches to environmental sustainability, which can all be used as building blocks for a new civilization.

You will need to overcome the false divide between science and religion and learn from religion the motivation to sacrifice in the short term for long-term benefits. Motivating change means reaching people's hearts as well as their minds, challenging their assumptions and empowering them to be open to change and work together for the betterment of their communities and society.

Since negative messages do not work, you will need a narrative of positive change and cooperation in a common cause, based on a spectrum of approaches and activating cooperative values rather than competitive values and stressing what we all have in common. You should relate solutions to each problem to the sources of happiness and the connections that we feel with others, creating communities of shared conviction. As you build your vision of the future, you will begin to see transformation as a journey of conviction and informed choice between desirable and catastrophic outcomes, which is based on the non-negotiable sacred values.

In sharing this vision with others, you will need to create mutual trust and be emotionally honest in talking about your hopes, fears and anxieties with moral consistency, recognizing your own role in the problems and affirming wider values across all the divisions between people. You will recognize people's feelings of grief and anxiety, mourn what is lost and value what remains. Acceptance, compassion, cooperation and empathy will produce very different outcomes than aggression, competition, blame and denial.[10]

This vision motivating positive change can show what the results of difficult efforts and short-term sacrifices can lead to. The gloomy, if not apocalyptic, environmental scenarios can be counterbalanced by visions of their role in the transition to a more peaceful world. You can envision a unified society that will be able to repair the damage and continue the onward march of civilization. An athlete supports endless hours of training and the often painful pushing of bodily limits, motivated by the ultimate satisfaction of winning, or at least of a race well run.

You can see now that environmental, economic and social catastrophes may represent a realistic probability in the short term, as it looks unlikely that our leaders will have the wisdom and courage to avoid them. Uniting before a common threat can be very beneficial in building a strong and resilient community and we all face common threats on this planet – as you have seen on your journey. Catastrophe is not only destructive, but forces a stronger spirit of cooperation.[11] The trials and suffering of World War II provided the impetus for the creation of the United Nations and the European Union. Unfortunately, it always seems to take such extreme events to break out of old paradigms and propel civilization another step forward. Therefore, it is necessary to follow your storyline or scenario of the future and be an example for others of responsible citizenship and a sustainable lifestyle to counteract the negative scenarios that may play out. You can become part of the constructive process of building a new global society on the other side of the catastrophe.

Furthermore, in such insecure times, where there is rising fear and mistrust, there is a natural tendency to want to return to the safety of old values and old ways of living, producing the driving force for the rise of fundamentalisms in many parts of the world, with accompanying intolerance, xenophobia and even terrorism. In visions of world citizenship and responsible consumption, you will need to offer a positive alternative, not looking to the past but to the future.

Global change is just one of the processes forcing us to recognize the reality of the oneness of humankind and our responsibility to make it happen. Globalization is opening up vast new potentials for the advancement of human civilization and many of our present difficulties are in fact the growing pains of the necessary transition.

You can have your own vision of possible sustainable societies that can result from implementing ethical values. If you are among the forward-looking generations of young people, you can imagine your own ideal future and take responsibility for your own life, creating in your mind and heart a goal worth sacrificing for. Your new generation of responsible consumers should not just be passively choosing what to buy from among what society offers, but actively preparing you to contribute in some way to the society you would like to live in. An ethical component in all forms of education can inspire you to become ethical leaders in your families, schools, communities, businesses, associations or governments. Or you may be at a more advanced stage in life, but prepared to break free from old ways of thinking and doing, ready to learn new ways and to contribute to a culture of change. In either case, the positive forces of ethical and spiritual commitment are the best motors for constructive change.

Scenarios of the future

One of the advantages of scenarios or storylines for the future is that they are not constraining. They describe possible futures extrapolated from a certain number of defining conditions and assumptions. They can warn us about things we should not do and suggest possible ways forward. If we do not like the kinds of futures some scenarios suggest, we are perfectly free to choose something else and make that happen. While we cannot directly affect processes in nature, like the sea level rise already under way with global warming, we have much more control over what people and human institutions do. As the old expression goes, 'we have found the enemy and it is us'. Furthermore, there is a natural process of change built into human life; the old people who often obstruct change will die off and the following generation will inevitably take over.

Below you will find visions of elements of the future from some of the peaks surrounding this valley. They take as a starting point the kind of human transformation described in the seven valleys of this metaphorical journey. As you join increasing numbers of people setting out on this voyage, many of the barriers that prevent humanity from fulfilling its true potential will recede. These glimpses of the possible future are compatible with the latest science and suggest where science can take us as we reach human maturity and leave behind the selfish and often violent practices of our collective adolescence.

Systems, knowledge and values

The United Nations Sustainable Development Goals already call for a systems perspective in which these ambitious goals are integrated and inseparable and must be achieved together. This will require new ways of structuring our institutions of governance, our academic disciplines and our economic and social structures. To be considered wise and learned in this new world will mean mastering all the major fields of knowledge, the sciences and arts and the essentials of religion and civilization, avoiding the traps of narrow specializations. Such knowledge will be accompanied by spiritual values of humility and service.

The environment

For too long we have taken our planet and its natural abundance for granted. It is, in fact, an amazing example and the only one we know of, where life has evolved to generate an unbelievable diversity of microbes, plants and animals – including us. At the moment we are rapidly destroying these beautifully balanced biological systems, but once we reconcile ourselves with nature we shall be able to restore much of the planet's productive capacity, rebuilding soils, recreating beautiful and productive landscapes and designing urban forest ecosystems in which we are fully integrated to produce the raw materials we need and to absorb our wastes in an organic unity. Agriculture, evolving far beyond the industrial monocultures of the moment, together with other renewable resources, will become the foundation of a sustainable society. As our knowledge of designing ecosystems advances, we may be able to raise the carrying capacity of the planet so that more people can benefit from our ever-advancing civilization.

Social justice

A united human race in all its diversity is now technically possible, as information and communications technologies have eliminated all barriers to world intercommunication. What we need is the social transformation to allow us to benefit from this new potential and to appreciate our diversity. The positive recognition of all those things that unite us will allow us to overcome national rivalries and hatreds, racial prejudice and religious strife. The nations of the world will agree on a world language and script to be taught in all the schools as an auxiliary to national languages, allowing a world literature and culture to emerge. The two great knowledge systems of science and religion, such potent forces in human life, will be reconciled and advance together. An independent press will express the diversified views and convictions of all of humanity.

We shall explore the optimal sizes of human communities for building strong social relationships, with a central house of worship as the spiritual heart of the community, surrounded by institutions of education and social welfare. Partisan politics will be abandoned in favour of consultative institutions of governance elected without nominations or campaigning, institutions whose only interest will be in serving the community.

Economy

The new world economy will uphold social justice, be altruistic and cooperative, create meaningful employment for all and eliminate poverty in the world. It will be managed at the global level to ensure that the planet's economic resources are organized, its raw materials are tapped and fully and responsibly utilized, its markets are coordinated and developed and its products distributed equitably. A single world currency will allow economic barriers and restrictions to be abolished. Destitution and gross accumulation of ownership will disappear and class distinctions will be removed. The end of economic and political warfare will allow resources to be directed to scientific research, new inventions and technical development; to increased productivity; to eliminating disease, improving physical health and extending human life; to refining the human brain; to the exploitation of all the planet's resources and sources of energy; and to stimulating the intellectual, moral and spiritual life of the entire human race.[12] An early priority will be to restore the environment and productive capacity of the planet from the damage we are causing now.

Each individual

Starting from the premise that our human purpose is to profit from this life to develop spiritually, to refine our character and to contribute to an ever-advancing civilization, we can imagine how different each individual life will be in this vision of the future. There will still be the universal challenges associated with the stages of life, from childhood, adolescence, adulthood, the decline in old age and ultimately death. But imagine how different it can be when infants are born into a loving family and every effort in the home, school and community helps each child discover the special gems within her or him and when training is provided for everyone to cultivate skills of service to the society and employment is guaranteed since work is a spiritual obligation.

Education will be life-long, to help everyone cultivate all the talents and abilities they possess; no one will be left behind. There will be no need for 'retirement', since opportunities for service will be adapted to each person's interests and capacities as they evolve through life. One could imagine a much more varied and balanced life, with time divided between work in agriculture or food production, technology or manufacturing, arts and crafts, education or health services, or administration and social contributions. While there would obviously still be specializations, we could leave behind rigid careers designed only to serve the economy rather than the fulfilment of the individual.

Community

All of these elements together can become the foundations for a spiritual civilization in which material progress enables the endless unfolding of the potentials of human consciousness. This is the real fruit of the unity of the human race, as our evolutionary progression opens the door to unimaginably rich social relationships and a blossoming of spiritual qualities in everyone. Bahá'u'lláh laid out this vision of the future in the nineteenth century and it is gradually unfolding, carried increasingly by young people and all generations. This is not just an imaginary utopia. The Bahá'í community is a learning community that does not have all the answers, but does have a set of spiritual principles that can help any community to find the answers to its problems and challenges. This is a reality that combines science and religion and the material and spiritual, with each reinforcing the other. It is intellectually coherent and emotionally satisfying.

Planetary vision

The unity of the human race will require a world commonwealth and federal system uniting all nations and peoples, while preserving state autonomy and the personal freedom and initiative of every individual. A world legislature, as the trustees of all of humanity, will control the resources of all the nations and make the laws necessary to regulate the lives, satisfy the needs and adjust the relationships of all races and peoples. A world executive, with the support of an international force, will apply the laws and safeguard the organic unity of the whole commonwealth. A world tribunal will settle all disputes that may arise within this universal system. This system, blending and embodying the ideals of East and West, liberated from war, with force acting as the servant of justice and unified materially and spiritually, is the goal of an emerging world civilization.[13]

Epilogue

You have now emerged from the seventh valley and can choose the peak to climb that best suits your interests, talents and circumstances. While this metaphorical journey is coming to an end, you still have your real, lifetime journey ahead. You are now armed for the constructive conquest of the planet. You can set out with a spirit of selfless service to work for the transformation of society, applying justice wherever you can. You can rise above the confusion and hopelessness of so many around you.

The future is in your hands. You can adopt new values and lifestyles and new forms of social organization. Where the institutions of society have become dysfunctional, as a result of the changing conditions in a globalized world, there is nothing except inertia to prevent you from replacing them with something better. The evolution of world society now requires a fundamental shift in perspective concerning the purpose of life, the nature of progress and the meaning of true happiness and wellbeing, as you have discovered on this journey. History is full of such transformations. Humanity is not forced to live forever with absolute national sovereignty, the unregulated neoliberal economy or the multinational corporation pursuing profit above all else. In the past, such transformations were brought about by war or revolution. Those options are still available. But the real challenge today is to ask if there are more civilized alternatives that do not require the recourse to violence and all the human suffering that accompanies it.

One thing that will give you hope is the new potential of information and communications technologies. Never before has it been possible for anyone, particularly the young generation, to access so much knowledge and wisdom, to network and exchange at scales unimaginable even a few decades ago and to organize and respond rapidly at unheard-of speeds. Previous transformations in human society were generally led by wise old men, establishing such innovations as the United Nations and the European Union. It is no longer necessary to wait for such leaders. Wisdom is available to anyone who looks for it. A new potential for positive change is opening up and the challenge for young people and everyone today is to prepare ourselves to grab this potential and run with it. As the Secretary-General of the United Nations has put it: 'Young people will be the torch bearers . . . the first truly globalized, interconnected and highly mobilized civil society, ready and able to serve as a participant, joint steward and powerful engine of change and transformation.'[1]

Starting in your local community

You have been through valleys of difficulties and challenges and climbed mountains of possibilities, but do not think that you have to go so far to take action. The best place to start to build a better world is within your own community. Invite others to come along with you and offer to accompany them on this journey that is both material and spiritual, finding local solutions to real problems and building in your community the capacity to take charge of its own improvement. The Bahá'ís have developed some excellent training materials prepared by the Ruhi Institute in Colombia, which are now used around the world in study circles, in case you want to learn how to teach children's classes in moral education or animate groups of pre-adolescent youth for service activities, as well as to advance in your own spiritual development. You can also hold devotional meetings where community members of all faiths and those of none can share some spiritual time together and build a deep sense of unity in the community. This approach is being used successfully all around the world in communities with many cultures, with people of all faith traditions, from remote tribal villages to the great cities, among rich and poor and across all the generations. You can probably find one near you.

You have armed yourself with a systems perspective that can help you to understand complex problems and a set of values to guide you through life. You now understand the roots of the environmental crisis and the solutions available. You have seen the importance of justice in addressing social problems and have the tools to launch yourself in social action. You are convinced of the need to transform a failing economic system and know of the steps you can take to start building an alternative from the bottom up. You have started on a life-long process of individual transformation in a spirit of humble learning. You appreciate the challenges associated with advancing your physical, intellectual and spiritual realities, learning to turn positively to the unknown in yourself, in others, in the natural world and reaching towards the Unknowable Essence of Absolute Perfection, that can inspire you to ever greater efforts in a path of service. And you have a clearer vision of the possible futures stretching out before you and the steps you can take to bring them a little closer. You are now full of hope, ready to face whatever challenges life may throw your way.

Bibliography

'Abdu'l-Bahá. 1957. *The Secret of Divine Civilization*. Wilmette, IL.: Bahá'í Publishing Trust, 1957.

Anello, Eloy. *A Framework for Good Governance in the Public Pharmaceutical Sector*, working draft for field testing and revision. Geneva: World Health Organization, 2008. Available at: www.who.int/entity/medicines/areas/policy/goodgovernance/GGMFramework2008-04-18.

Bahá'í International Community. *The Prosperity of Humankind*, statement prepared by the Bahá'í International Community Office of Public Information, Haifa, Israel. London: Bahá'í Publishing Trust, 1995. Available at: http://iefworld.org/bicpros.htm.

— *Valuing Spirituality in Development: Initial Considerations Regarding the Creation of Spiritually Based Indicators for Development*, concept paper written for the World Faiths and Development Dialogue, Lambeth Palace, London, 18–19 February 1998. London: Bahá'í Publishing Trust, 1998. Available at: http://iefworld.org/bicvsid.htm.

Bahá'u'lláh. *Gleanings from the Writings of Bahá'u'lláh*, §82. Wilmette, IL.: Bahá'í Publishing Trust, 1976.

— *The Hidden Words of Bahá'u'lláh*, Wilmette, IL.: Bahá'í Publishing Trust, 1954.

— *The Seven Valleys and the Four Valleys*. Trans. M. Gail with A-K. Khan. Wilmette, IL: Bahá'í Publishing Trust, rev. ed. 1975.

— *The Summons of the Lord of Hosts*. Haifa: Bahá'í World Centre, 2002.

Beddington, John. Speech at GovNet SDUK09, 2009. Available at: http://www.govnet.co.uk/news/govnet/professor-sir-john-beddingtons-speech-at-sduk-09.

Beinhocker, Eric D. *The Origin of Wealth: Evolution, Complexity, and the*

Radical Remaking of Economics. Cambridge: Harvard Business School Press and London: Random House Business Books, 2006.

Besson, Jacques. 'L'impact de la spiritualité dans le domaine de l'addiction est avéré', interview by Laurent Nicolet in *Migros Magazine*, no. 31, 2 August 2016, pp. 34–7.

Blanke, Jennifer; Loades, Emma. 'Capturing the state of country competitiveness with the Executive Opinion Survey', in Michael E. Porter, Klaus Schwab, Xavier Sala-i-Martin and Augusto Lopez-Claros: *The Global Competitiveness Report 2004–2005*, World Economic Forum (Houndsmill, UK and New York: Palgrave Macmillan, 2004), pp. 199–208.

Braconnier, Alain. *Optimiste*. Paris: Odile Jacob, 2014.

— 'L'optimisme donne du sense à l'existence', interview in *Migros Magazine*, no. 10, 3 March 2014, pp. 20–23.

Brooks, Michael. 'If information . . . then universe', and 'Does consciousness create reality', in *New Scientist*, special issue:, 29 September 2012, pp. 41 and 42–3.

Brown, Paul. 'Climate change threatens global financial crash', in *Climate News Network*, 2 October 2015. Available at: http://climatenewsnetwork.net/climate-change-threatens-global-financial-crash/?utm_source=Climate+News+Network&utm_campaign=1bd3656d6b-Carney_s_crash_warning10_2_2015&utm_medium=email&utm_term=0_1198ea8936-1bd3656d6b-38775505.

Buchanan, Mark. 'No end to the multiverses', review of Max Tegmar, 'Our mathematical universe: My quest for the ultimate nature of reality', in *New Scientist*, vol. 221, no. 2952, 18 January 2014, pp. 46–7.

Burford, Gemma; Hoover, E.; Velasco, I.; Janoušková, I.; Jimenez, A.; Piggot, G.; Podger, D.; M. K. Harder. 'Bringing the "missing pillar" into Sustainable Development Goals: Towards intersubjective values-based indicators', in *Sustainability*, vol. 5 (2013), pp. 3035–59. doi:10.3390/su5073035 http://www.mdpi.com/2071-1050/5/7/3035

Carrington, Damian. 'David Attenborough: Collapse of civilisation is on the horizon', in *The Guardian*, 3 December 2018. Available at: https://www.theguardian.com/environment/2018/dec/03/david-attenborough-collapse-civilisation-on-horizon-un-climate-summit.

Carroll, Sean. 'It's mind-blowing what our puny brains can do', interview in *New Scientist*, vol. 230, no. 3069, 16 April 2016, pp. 28–9.

Dahl, Arthur Lyon. *The Eco Principle: Ecology and Economics in Symbiosis.* London: Zed Books and Oxford: George Ronald, 1996.

— 'The competitive edge in environmental responsibility', in Michael E. Porter, Klaus Schwab, Xavier Sala-i-Martin and Augusto Lopez-Claros: *The Global Competitiveness Report 2004–2005,* World Economic Forum (Houndsmill, UK and New York: Palgrave Macmillan, 2004), pp. 103–110.

— *Preventing Overshoot and Collapse: Managing the Earth's Resources,* paper prepared on the introductory theme of the 2008 UNEP/University of Geneva/Graduate Institute Environmental Diplomacy Course, August 2008. Available at: http://iefworld.org/ddahlo8d.htm.

— *Financial Crisis and the Green Economy,* paper prepared on the introductory theme of the 2009 UNEP/University of Geneva/Graduate Institute Environmental Diplomacy Course, August 2009. Available at: http://iefworld.org/ddahlo9d.html.

— 'Interstate collaboration for human security: The lessons from Copenhagen', in Takehiro Togo and Negoslav P. Ostojic (eds): *National and Inter-ethnic Reconciliation, Religious Tolerance and Human Security in the Balkans: Human Security Concept Implementation,* Proceedings of the 6th ECPD International Conference, Brioni Islands, Croatia, 28–29 October 2010 (Belgrade: European Center for Peace and Development, 2011), pp. 59–61. Available at: http://iefworld.org/ddahl10c.

Diamond, Jared. *Collapse: How Societies Choose to Fail or Survive.* London: Allen Lane and New York: Viking Penguin, 2005.

Earth Charter Initiative. *The Earth Charter.* San José, Costa Rica, Earth Charter Initiative, 2000. Available at: http://www.earthcharter.org/.

Esslemont, J. E. *Bahá'u'lláh and the New Era* (1923). Wilmette IL: Bahá'í Publishing Trust, 1980.

Falomir-Pichastor, Joan. 2016. 'Nous réussissons à faire tout et son contraire sans percevoir d'incohérence', interview by Laurent Nicolet in *Migros Magazine,* no. 25, 20 June 2016, pp. 40–43.

Gefter, Amanda. 'Is everything made of numbers?', in *New Scientist,* special issue: *Reality,* 29 September 2012, pp. 38–9.

Global Agenda Council on Climate Change. 2009. *Shaping an Opportunity Out of Crisis,* a message to participants in the World Economic Forum Annual Meeting 2009 from Members of the Global Agenda Council on Climate Change. Geneva: World Economic Forum, January 2009. Available at: http://www.weforum.org/pdf/GAC/selectedcontributions/Global Agenda Council on Climate Change.pdf.

Global Commission on the Economy and Climate. 2014. *Better Growth, Better Climate: The New Climate Economy Report*. Washington DC, Global Commission on the Economy and Climate, 2014.

Hammond, Allen. *Which World? Scenarios for the 21st Century: Global Destinies, Regional Choices*. Washington DC: Island Press, 1998.

Hanley, Paul. *Eleven*. Victoria, BC, Canada: Friesen Press, 2014.

Hibbing, John; Smith, Kevin. 'We are what we vote', in *New Scientist*, vol. 226, no. 3015, 4 April 2014, pp. 24–25, based on John Hibbing, Kevin Smith and John R. Alford: *Predisposed: Liberals, Conservatives, and the Biology of Political Differences* (London: Routledge, 2013).

Holderness, Mike. 'How do we know?', in *New Scientist*, special issue: *Reality*, 29 September 2012, p. 45.

Homer-Dixon, Thomas. *The Upside of Down: Catastrophe, Creativity, and the Renewal of Civilization*. Toronto: Vintage Canada, 2006.

Intergovernmental Panel on Climate Change (IPCC). *Climate Change 2014: Synthesis Report*, Approved Summary for Policymakers. Geneva: IPCC, 2014. Available at: http://www.ipcc.ch/pdf/assessment-report/ar5/syr/SYR_AR5_SPM.pdf.

— *Global Warming of 1.5°C (SR15): Special Report*, Summary for Policymakers, October 2018. Geneva: IPCC, 2018.

International Institute for Strategic Studies. *Strategic Survey 2007*. London: IISS, 2007.

Jacobson, Mark Z.; Delucchi, Mark A.. 2011. 'Providing all global energy with wind, water, and solar power, Part I: Technologies, energy resources, quantities and areas of infrastructure, and materials', in *Energy Policy*, vol. 39 (2011), pp. 1154–1169. doi:10.1016/j.enpol.2010.11.040.

Jamieson, Valerie. 'The bedrock of it all', in *New Scientist*, special issue: *Reality*, 29 September 2012, p. 36.

Jamison, Rob. 'The blunders that led to catastrophe', in *New Scientist*, 27 September 2008, pp. 8–9.

Kanai, R.; Feilden, T.; Firth, C; Rees, G. 2011. 'Political orientations are correlated with brain structure in young adults', in *Current Biology*, vol.

21, no. 8 (26 Apr. 2011), pp. 677–680. doi: 10.1016/j.cub.2011.03.017.

Korowicz, David. *Tipping Point: Near-term Systemic Implications of a Peak in Global Oil Production, An Outline Review.* Dublin, Ireland: Foundation for the Economics of Sustainability (Feasta), 2010. Available at: http://www.feasta.org/documents/risk_resilience/Tipping_Point.pdf.

Lawton, Graham. 'Beyond belief'. In *New Scientist*, vol. 226, no. 3015, 4 April 2015, pp. 28–33.

Lopez-Claros, Augusto. Letter to the Editor of the *Financial Times*, 4 December 2008.

MacKenzie, Debora. 'The end of civilization', in *New Scientist*, 5 April 2008, pp. 28–31.

— 'Are we doomed? The very nature of civilization may make its demise inevitable', in *New Scientist*, 5 April 2008, pp. 32–5.

— 'Doomsday Book', in *New Scientist*, 7 January 2012, pp. 38–41.

— 'On the road again', in *New Scientist*, vol. 230, no. 3068, 9 April 2016, pp. 29–37.

Marshall, George. *Don't Even Think About It: Why Our Brains are Wired to Ignore Climate Change.* London and New York: Bloomsbury, 2014.

McKibben, Bill. 'Global warming's terrifying new math', in *Rolling Stone*, 2 August 2012. Available at: http://www.rollingstone.com/politics/news/global-warmings-terrifying-new-math-20120719.

Meadows, D. H.; Meadows, D. L.; Randers, J.; Behrens W. W. III. *The Limits to Growth*, A Report for the Club of Rome's Project on the Predicament of Mankind. New York: Universe Books, 1972.

—; —; —. *Beyond the Limits: Confronting Global Collapse, Envisioning a Sustainable Future.* White River Junction, Vermont: Chelsea Green, 1992.

—; —; —. 2004. *Limits to Growth: The 30-Year Update.* White River Junction, Vermont: Chelsea Green, 2004.

Montgomery, David R. *Dirt: The Erosion of Civilizations.* Berkeley: University of California Press, 2007.

New Scientist. 'The collapse of civilization: It's more precarious than we realized', cover story, 5 April 2008; 'For whom the bell tolls', ibid. p. 5.

— Editorial, in vol. 226, no. 3015, 4 April 2015.

— Special issue: *Reality*, 29 September 2012, pp. 35–47, including: Jan

Westerhoff, 'Defining reality'. p. 35, and 'Is matter real?', pp. 37–46; Valerie Jamieson, 'The bedrock of it all'. p. 36; Amanda Gefter, 'Is everything made of numbers?', pp. 38–9; Michael Brooks, 'If information . . . then universe'. p. 41, and 'Does consciousness create reality?' pp. 42–3; Mike Holderness, 'How do we know?' p. 45; Richard Webb, 'The future of reality'. p. 47.

Nowak, Martin A.; Highfield, Roger. *SuperCooperators: Altruism, Evolution, and Why We Need Each Other to Succeed.* New York: Free Press, 2011.

OSED (Office of Social and Economic Development). *Social Action*, paper prepared by the Office of Social and Economic Development at the Bahá'í World Centre, Haifa, Israel, 26 November 2012.

PERL (Partnership for Education and Research about Responsible Living). *Discovering What Matters: A Journey of Thinking and Feeling*, PERL Values-based Learning Student Toolkit. Hamar, Norway: PERL, 2014.

— *Measuring What Matters: Values-based Indicators*, PERL Values-based Learning Methods Toolkit. Hamar, Norway: PERL, 2014.

— *Building a Shared Vision: A Toolkit for Schools*, PERL Values-based Learning Staff Toolkit. Hamar, Norway: PERL, 2014.

Pinker, Steven. *The Blank Slate* (2002), quoted in Edward O. Wilson: *The Social Conquest of Earth* (New York: Liveright, 2012), p. 248.

Podger, Dimity; Velasco, I.; Luna, C. A.; Burford, G.; Harder, M. K. 2013. 'Can values be measured? Significant contributions from a small civil society organisation through action research.', in *Action Research Journal*, vol. 11 (2013), no. 1, pp. 8–30. doi: 10.1177/1476750312467833.

—; Hoover, E.; Burford, G.; Hak, T.; Harder, M. K. 2015. 'Revealing values in a complex environmental program: A scaling up of values based in dicators', in *Journal of Cleaner Production*, August 2015. doi: 10.1016/j. jclepro.2015.08.034.

Pope Francis. *Laudato Si': On Care for our Common Home.* Encyclical, 18 June 2015). Available at: http://w2.vatican.va/content/francesco/en/encyclicals/documents/papa-francesco_20150524_enciclica-laudato-si. html.

Randers, Jorgen. *2052: A Global Forecast for the Next Forty Years*, report to The Club of Rome. White River Junction, Vermont: Chelsea Green Publishing, 2012.

Raskin, Paul. *Journey to Earthland: The Great Transition to Planetary Civiliza-*

tion. Boston: Tellus Institute, 2016. Available at : http://www.tellus.org/ tellus/publication/journey-to-earthland.

—; Banuri, T.; Gallopin, G.; Gutman, P.; Hammond, A.; Kates, R.; Swart, R. *Great Transition: The Promise and Lure of the Times Ahead*. Boston: Stockholm Environment Institute, and Global Scenario Group, 2002.

Rawls, John. *A Theory of Justice*. Cambridge: Harvard University Press, rev. ed. 1999.

Rockström, J.; Steffen, W.; Noone, K.; Persson, A. et al. 2009. 'Planetary boundaries: Exploring the safe operating space for humanity', in *Ecology and Society*, vol. 14 (2009), no. 2, p. 32. Available at: http://www.ecologyandsociety.org/vol14/iss2/art32/.

Sachs, Jeffrey D. 'Restoring virtue ethics in the quest for happiness', Chapter 5 in John Helliwell, Richard Layard and Jeffrey Sachs (eds): *World Happiness Report 2013* (New York: Sustainable Development Solutions Network and Earth Institute, Columbia University, 2013), pp. 80–97.

Saucier, Gerald. *Journal of Personality and Social Psychology*, vol. 104 (2015), p. 921, quoted in Graham Lawton: 'Beyond belief', in *New Scientist*, vol. 226, no. 3015, 4 April 2015, pp. 28–33.

Seager, Ashley. 'Torrent of bad news ends hope of "quick" recession', in *The Guardian Weekly*, 27 February–5 March 2009, pp. 1–2.

Shoghi Effendi. *The World Order of Baháʼuʼlláh* (1938). Wilmette, IL.: Baháʼí Publishing Trust, 2nd rev. ed. 1974.

Sichko, S.; Borelli, J. L., Smiley, P. A., Goldstein, A.; Rasmussen, H. F. 2011. 'Child and maternal attachment predict school-aged children's psychobiological convergence,' in *Developmental Psychobiology*, vol. 60 (2018), no. 1, pp. 1–14. doi: 10 .1002/dev.21748.

Smith, Adam. *The Wealth of Nations* (1776), Book IV, Chapter II, paragraph IX. London: Penguin/Random House, 2003.

Spiegel Online. 2009. 'Can countries really go bankrupt?' 30 January 2009. Available at:http://www.spiegel.de/international/world/0,1518,604523,00. html.

Steffen, W.; Richardson, K.; Rockström, J.; Cornell, S. E. et al. 'Planetary boundaries: Guiding human development on a changing planet' in *Science*, vol. 347, no. 6223 (13 Feb. 2015). doi: 10.1126/science.1259855.

Stern, Nicholas. *The Economics of Climate Change*. London: Her Majesty's Treasury, 2006; Cambridge: Cambridge University Press, 2007. Available

at: http://webarchive.nationalarchives.gov.uk/20080910140413/http://
www.hm-treasury.gov.uk/independent_reviews/stern_review_econom-
ics_climate_change/sternreview_index.cfm.

Thomas, Chris. 2014. 'Evolutionary explosion', interview with Fred Pearce
in *New Scientist*, vol. 221, no. 2951, 11 Jan. 2014, pp. 28–29.

Turchin, Peter. *War and Peace and War: The Rise and Fall of Empires*. New
York: Plume Books (Penguin), 2006.

— 'Political instability may be a contributor in the coming decade', in *Nature*,
vol. 463, issue 7281 (4 February 2010), p. 608. doi:10.1038/463608a.

— *Ultra Society: How 10,000 Years of War Made Humans the Greatest Cooperators on Earth*. Chaplin, Connecticut: Beresta Books, 2016.

UNDP (United Nations Development Programme). *Human Development
Report 2007/2008: Fighting Climate Change: Human Solidarity in a
Divided World*. New York: Palgrave Macmillan, for United Nations De-
velopment Programme, 2007.

UNEP (United Nations Environment Programme). *Global Crises: National
Chaos?*, GC/GMEF 25th Regular Session, Ministerial Consultations,
Background Paper on Theme I, Globalization and the Environment.
UNEP/GC.25/16, 2009.

United Nations. *Universal Declaration of Human Rights* (New York, United
Nations, 1948). Available at: http://www.unhchr.ch/udhr/index.htm.

— Declaration of the United Nations Conference on the Human Environment (Stockholm Declaration). Report of the United Nations Conference
on the Human Environment, Stockholm, 5–16 June 1972.

—*Rio Declaration on Environment and Development*, United Nations General
Assembly A/CONF. 11/26 Annex 1, New York, United Nations, 1992.
Available at: http://www.un.org/documents/ga/conf151/aconf15126-
1annex1.htm.

— *The Road to Dignity by 2030: Ending Poverty, Transforming All Lives
and Protecting the Planet*, Synthesis Report of the Secretary-General on
the Post-2015 Agenda, Document A/69/700, 4 December 2014. New
York: United Nations, 2014. http://www.un.org/ga/search/view_doc.
asp?symbol=A/69/700&Lang=E.

— 2015. *Transforming Our World: The 2030 Agenda for Sustainable Development*, Outcome document of the Summit for the adoption of the

Post-2015 Development Agenda, New York, 25–27 September 2015. A/70/L.1. New York: United Nations, 2015. Available at: http://www. un.org/ga/search/view_doc.asp?symbol=A/70/L.1&Lang=E .

The Universal House of Justice. *One Common Faith.* Wilmette, IL.: Bahá'í Publishing Trust, 2005.

— 'To the Bahá'ís of the World', Riḍván 2012.

— 'To the Bahá'ís of the World', 8 February 2013, Bahá'í World Centre, Haifa.

— 'To the believers in the Cradle of the Faith', 2 April 2010. Bahá'í World Centre, Haifa.

— 'To the Conference of the Continental Boards of Counsellors', 28 December 2010. Bahá'í World Centre, Haifa.

Vidal, John. *The Guardian Weekly,* 9–15 February 2007, Energy supplement, p. 3.

Webb, Richard. 'The future of reality'. in *New Scientist,* special issue: *Reality,* 29 September 2012, p. 47.

Westerhoff, Jan. 2012. 'Defining reality'. And 'Is matter real?', in *New Scientist,* special issue: *Reality,* 29 September 2012, pp. 35 and 37–46.

Wilson, Edward O. *The Social Conquest of Earth.* New York: Liveright, 2012. Introduction

Notes and References

Introduction

1 The concept of Seven Valleys has origins in the 12th-century Sufi poet Farid al-Din Attar's *Conference of the Birds*. It was taken up by Bahá'u'lláh in his mystical work, *The Seven Valleys*, addressed to a Sufi Shaykh.

1 Valley of Lost Souls

1 Besson, 'L'impact de la spiritualité dans le domaine de l'addiction est avéré'.

2 ibid.

3 If you want to explore this in more detail, see Dahl, *The Eco Principle: Ecology and Economics in Symbiosis* (1996).

4 Wilson, *The Social Conquest of Earth*, Chapter 24: 'The origins of morality and honor'.

5 Turchin, *War and Peace and War: The Rise and Fall of Empires*.

6 Anello, *A Framework for Good Governance in the Public Pharmaceutical Sector*, p. 45.

7 Rawls, *A Theory of Justice*, pp. 3–4.

8 Marshall, *Don't Even Think About It: Why Our Brains are Wired to Ignore Climate Change*.

9 Burford et al., 'Bringing the "missing pillar" into Sustainable Development Goals: Towards intersubjective values-based indicators'; Podger et al., 'Revealing values in a complex environmental program: A scaling up of values-based indicators'.

10 PERL, *Discovering What Matters: A Journey of Thinking and Feeling*; *Measuring What Matters: Values-based Indicators*; *Building a Shared Vision: A Toolkit for Schools*.

2 The Valley of Environmental Crises

1 Meadows et al., *The Limits to Growth*.
2 Montgomery, *Dirt: The Erosion of Civilizations*.
3 Meadows et al., *Beyond the Limits: Confronting Global Collapse, Envisioning a Sustainable Future* (1992).
4 Meadows et al. *Limits to Growth: The 30-Year Update* (2004).
5 MacKenzie, 'Doomsday Book'.
6 Diamond, *Collapse: How Societies Choose to Fail or Survive*; Homer-Dixon, *The Upside of Down: Catastrophe, Creativity, and the Renewal of Civilization*.
7 Turchin, *War and Peace and War: The Rise and Fall of Empires*.
8 Turchin, 'Political instability may be a contributor in the coming decade'.
9 Turchin, *War and Peace and War: The Rise and Fall of Empires*.
10 Hammond, *Which World? Scenarios for the 21st Century: Global Destinies, Regional Choices*.
11 Homer-Dixon, *The Upside of Down: Catastrophe, Creativity, and the Renewal of Civilization*; Hanley, *Eleven*.
12 Raskin et al. *Great Transition: The Promise and Lure of the Times Ahead*; Raskin, *Journey to Earthland: The Great Transition to Planetary Civilization*. See also http://www.GTInitiative.org.
13 *New Scientist*, 'The collapse of civilization: It's more precarious than we realized', cover story, 5 April 2008; Editorial, 'For whom the bell tolls', ibid. p. 5; MacKenzie, 'The end of civilization', ibid. pp. 28–31; MacKenzie, 'Are we doomed? The very nature of civilization may make its demise inevitable', ibid. pp. 32–5.
14 Beddington, speech at GovNet SDUK09 (2009).
15 Dahl, *Preventing Overshoot and Collapse: Managing the Earth's Resources*.
16 Stern, *The Economics of Climate Change*.
17 Vidal, *The Guardian Weekly*, 9–15 February 2007, Energy supplement, p. 3.
18 Letter from the Universal House of Justice to an individual, 25 December 2014.
19 Dahl, *The Eco Principle: Ecology and Economics in Symbiosis*.
20 Rockström et al., 'Planetary boundaries: Exploring the safe operating space for humanity'; Steffen et al., 'Planetary boundaries: Guiding human development on a changing planet'.
21 Dahl, 'Interstate collaboration for human security: The lessons from Copenhagen'.
22 McKibben, 'Global warming's terrifying new math'.
23 Intergovernmental Panel on Climate Change (IPCC), *Climate Change 2014: Synthesis Report*, Approved Summary for Policymakers; IPCC, *Global Warming of 1.5°C (SR15)*, Special Report, Summary for Policymakers, October 2018.

24 International Institute for Strategic Studies, *Strategic Survey 2007*.
25 Attenborough, quoted in Carrington, 'David Attenborough: Collapse of civilsation is on the horizon', 3 December 2018.
26 Stern, *The Economics of Climate Change*.
27 Jacobson and Delucchi, 'Providing all global energy with wind, water, and solar power, Part I: Technologies, energy resources, quantities and areas of infrastructure, and materials'.
28 Thomas, 'Evolutionary explosion'.
29 Montgomery, *Dirt: The Erosion of Civilizations*.
30 Hanley, *Eleven*.
31 Dahl, *The Eco Principle: Ecology and Economics in Symbiosis*.
32 Marshall, *Don't Even Think About It: Why Our Brains are Wired to Ignore Climate Change*.
33 ibid.

3 The Valley of Social Illnesses

1 MacKenzie, 'On the road again'.
2 Turchin, *War and Peace and War: The Rise and Fall of Empires*.
3 Turchin, *Ultra Society: How 10,000 Years of War Made Humans the Greatest Cooperators on Earth*.
4 Wilson, *The Social Conquest of Earth*.
5 Pinker, *The Blank Slate* (2002), quoted in Wilson, *The Social Conquest of Earth*.
6 Nowak and Highfield, *SuperCooperators: Altruism, Evolution, and Why We Need Each Other to Succeed*.
7 Sachs, 'Restoring virtue ethics in the quest for happiness', in Helliwell, Layard and Sachs (eds): *World Happiness Report 2013*.
8 OSED, *Social Action*.
9 United Nations, *The Road to Dignity by 2030: Ending Poverty, Transforming All Lives and Protecting the Planet*.
10 OSED, *Social Action*.
11 ibid.
12 ibid.
13 ibid.

4 The Valley of Economic Crisis

1 Dahl, *Financial Crisis and the Green Economy*.
2 Jamison, 'The blunders that led to catastrophe'.
3 Jean-Claude Trichet, quoted in Seager, 'Torrent of bad news ends hope of "quick" recession', pp. 1–2.
4 Lopez-Claros, Letter to the Editor of the *Financial Times*, 4 December 2008.
5 *Spiegel Online*, 'Can countries really go bankrupt?'.

6 Mark Carney, quoted in Brown, 'Climate change threatens global financial crash'.
7 Korowicz, *Tipping Point: Near-term Systemic Implications of a Peak in Global Oil Production, An Outline Review.*
8 Randers, *2052: A Global Forecast for the Next Forty Years.*
9 ibid.
10 ibid.
11 ibid.
12 UNEP, *Global Crises: National Chaos?*
13 Global Agenda Council on Climate Change, *Shaping an Opportunity Out of Crisis.*
14 Global Commission on the Economy and Climate, *Better Growth, Better Climate: The New Climate Economy Report.*
15 Blanke and Loades, 'Capturing the state of country competitiveness with the Executive Opinion Survey', in Porter, Schwab, Sala-i-Martin and Lopez-Claros, *The Global Competitiveness Report 2004–2005.*
16 Dahl, 'The competitive edge in environmental responsibility', in Porter, Schwab, Sala-i-Martin and Lopez-Claros, *The Global Competitiveness Report 2004–2005.*
17 United Nations, *Universal Declaration of Human Rights.*
18 United Nations, *Declaration of the United Nations Conference on the Human Environment.*
19 United Nations, *Rio Declaration on Environment and Development.*
20 Earth Charter Initiative, *The Earth Charter.*
21 Smith, *The Wealth of Nations*, Book IV, Ch. 1.
22 Bahá'í International Community, *Valuing Spirituality in Development: Initial Considerations Regarding the Creation of Spiritually Based Indicators for Development.*
23 UNDP, *Human Development Report 2007/2008: Fighting Climate Change: Human Solidarity in a Divided World*, p. 68.
24 Pope Francis, *Laudato Si': On Care for our Common Home.*
25 Universal House of Justice, *One Common Faith.*
26 Bahá'u'lláh, *The Summons of the Lord of Hosts*, p. 193.
27 Bahá'í International Community, *Valuing Spirituality in Development: Initial Considerations Regarding the Creation of Spiritually Based Indicators for Development.*
28 ibid.
29 ibid.
30 Bahá'í International Community, *The Prosperity of Humankind.*
31 ibid.
32 Beinhocker, *The Origin of Wealth: Evolution, Complexity, and the Radical Remaking of Economics.*
33 Burford et al., 'Bringing the "missing pillar" into Sustainable

Development Goals: Towards intersubjective values-based indicators';
Podger et al., 'Can values be measured? Significant contributions from
a small civil society organization through action research'.

34 The Universal House of Justice, 'To the believers in the Cradle of the
Faith', 2 April 2010.

35 ibid.

36 United Nations, *The Road to Dignity by 2030: Ending Poverty, Transforming All Lives and Protecting the Planet.*

37 United Nation, *Transforming Our World: The 2030 Agenda for Sustainable Development.*

38 ibid.

39 ibid.

5 The Valley of Individual Discovery

1 The Universal House of Justice, 'To the Bahá'ís of the World', Riḍván
2012.

2 Braconnier, *Optimiste*; Braconnier, 'L'optimisme donne du sense à
l'existence'.

3 Kanai et al., 'Political orientations are correlated with brain structure in
young adults'.

4 Sichko, Borelli, Smiley, Goldstein and Rasmussen, 'Child and maternal attachment predict school-aged children's psychobiological
convergence'.

5 Hibbing and Smith, 'We are what we vote', 4 April 2014, based on
Hibbing, Smith and Alford, *Predisposed: Liberals, Conservatives, and the
Biology of Political Differences.*

6 Braconnier, *Optimiste*; Braconnier, 'L'optimisme donne du sense à
l'existence'.

7 Turchin, *War and Peace and War: The Rise and Fall of Empires*; Turchin,
*Ultra Society: How 10,000 Years of War Made Humans the Greatest Coop-
erators on Earth.*

8 Lawton, 'Beyond belief'.

9 ibid.

10 ibid.

11 ibid.

12 ibid.

13 Falomir-Pichastor, 'Nous réussissons à faire tout et son contraire sans
percevoir d'incohérence'.

14 ibid.

15 Saucier, in *Journal of Personality and Social Psychology*, quoted in
Lawton, 'Beyond belief'.

16 Lawton, 'Beyond belief'.

17 ibid.

18 Bahá'u'lláh, quoted in Esslemont, *Bahá'u'lláh and the New Era*, Ch. 3, p. 35.

19 'Abdu'l-Bahá, *The Secret of Divine Civilization*, p. 59.

6 The Valley of Multiple Realities

1 See *New Scientist*, Special issue: *Reality*, 29 September 2012, various authors.

2 ibid.

3 Westerhoff, 'Is matter real?', ibid.

4 Gefter, 'Is everything made of numbers?', ibid.

5 Brooks, 'If information . . . then universe', ibid.

6 Brooks, 'Does consciousness create reality?', ibid.

7 Holderness, 'How do we know?', ibid.

8 Webb, 'The future of reality', ibid.

9 Carroll, 'It's mind-blowing what our puny brains can do'.

10 Buchanan, 'No end to the multiverses'.

11 OSED, *Social Action*.

12 Bahá'u'lláh, *Hidden Words*, Arabic no. 66.

13 ibid. Arabic no. 5.

14 The concept of Seven Valleys has origins in the 12th-century Sufi poet Farid al-Din Attar's *Conference of the Birds*. It was taken up by Bahá'u'lláh in his mystical work *The Seven Valleys*, addressed to a Sufi Shaykh.

15 Bahá'u'lláh, *Gleanings from the Writings of Bahá'u'lláh* LXXXII.

16 ibid.

7 The Valley of Visions of the Future

1 Turchin, 'Political instability may be a contributor in the coming decade'.

2 ibid.

3 The Universal House of Justice, 'To the believers in the Cradle of the Faith', 2 April 2010

4 Paraphrase from the Universal House of Justice, 'To the believers in the Cradle of the Faith', 2 April 2010.

5 Paraphrase from the Universal House of Justice, 'To the Conference of the Continental Boards of Counsellors', 28 December 2010, Bahá'í World Centre, Haifa.

6 Inspired by the Universal House of Justice, 'To the Bahá'ís of the World', 8 February 2013.

7 ibid.

8 Paraphrase from the Universal House of Justice, 'To the believers in the Cradle of the Faith', 2 April 2010.

9 ibid.

10 Marshall, *Don't Even Think About It: Why Our Brains are Wired to Ignore Climate Change.*

11 Turchin, *Ultra Society: How 10,000 Years of War Made Humans the Greatest Cooperators on Earth.*

12 After Shoghi Effendi, *The World Order of Bahá'u'lláh.*

13 ibid.

Epilogue

1 United Nations, *The Road to Dignity by 2030: Ending Poverty, Transforming All Lives and Protecting the Planet.*

About the author

Dr Arthur Lyon Dahl is President of the International Environment Forum and a retired senior official of UN Environment, with over 50 years experience in sustainability, international environmental assessment and governance, and systems science, including more than 20 years in intergovernmental organizations. A biologist by training (Stanford University and University of California, Santa Barbara), specializing on small islands and coral reefs, he lived and worked for many years in the Pacific Islands and Africa before settling in Geneva, Switzerland.

A member of the Bahá'í Faith, he chose science as his field of service to humanity and the planet, and has always sought coherence between his values and his work. He has published over 200 papers and books including: *Unless and Until: A Bahá'í Focus on the Environment* and *The Eco Principle: Ecology and Economics in Symbiosis*.

Lightning Source UK Ltd.
Milton Keynes UK
UKHW010613181119
353752UK00001B/131/P